CASE STUDIES IN VETERINARY IMMUNOLOGY

CASE STUDIES IN VETERINARY IMMUNOLOGY

Laurel J. Gershwin
University of California, Davis

Garland Science
Taylor & Francis Group
NEW YORK AND LONDON

Vice President:	Denise Schanck
Senior Editorial Assistant:	Katie Laurentiev
Senior Digital Project Editor:	Natasha Wolfe
Typesetter:	Thomson Digital
Copy Editor:	Josephine Hargreaves
Proofreader:	Susan Wood
Cover and Text Design:	Matthew McClements, Blink Studio, Ltd.
Illustrator:	Nigel Orme
Indexer:	Simon Yapp at Indexing Specialists

ISBN 978-0-8153-4447-6

Library of Congress Cataloging-in-Publication Data

Names: Gershwin, Laurel J., editor.
Title: Case studies in veterinary immunology / [edited by] Laurel Gershwin.
Description: New York, NY : Garland Science, [2017]
Identifiers: LCCN 2016058716 | ISBN 9780815344476
Subjects: | MESH: Animal Diseases—immunology | Immune System Diseases—veterinary | Case Reports
Classification: LCC SF757.2 | NLM SF 757.2 | DDC 636.0896/079—dc23
LC record available at https://lccn.loc.gov/2016058716

Published by Garland Science, Taylor & Francis Group, LLC, an informa business, 711 Third Avenue, New York, NY 10017, USA, and
3 Park Square, Milton Park, Abingdon, OX14 4RN, UK.
Printed in the United States of America
15 14 13 12 11 10 9 8 7 6 5 4 3 2 1

SUSTAINABLE FORESTRY INITIATIVE

Certified Chain of Custody
Promoting Sustainable Forestry

www.sfiprogram.org
SFI-01681

Visit our website at www.garlandscience.com

PREFACE

This book was inspired by Raif Geha and Luigi Notarangelo's excellent clinical guide to human immunological diseases, *Case Studies in Immunology: A Clinical Companion*, also published by Garland Science. It was developed as a resource for students of veterinary immunology, who sometimes struggle with the clinical relevance of complex immunological mechanisms, and my hope is that it will serve as a valuable supplement and companion to a variety of immunology textbooks and courses. It should also be very useful to residents preparing for board examinations, to graduate veterinarians, and to anyone with an interest in comparative immunology.

The book is arranged according to mechanisms of disease causation, and includes cases of both primary and secondary immunodeficiency, auto-immunity, and those that can be classified as one or more of the four types of hypersensitivity reaction described by Gell and Coombs. Each case includes a workup, a list of differential diagnoses, pertinent diagnostic laboratory data, and treatment options. Comparative medicine considerations at the end of each case engage with the similarities and differences that occur among different species with the same or similar medical conditions, including humans. End-of-case questions and answers for student assessment, and a list of further reading for those who wish to delve deeper into the literature, have also been included. Appendices that describe common vaccination protocols for domestic animal species, as well as some of the more common diagnostic tests performed in a clinical immunology laboratory, serve as a helpful clinical reference.

Case Studies in Veterinary Immunology contains cases drawn from a wide variety of species—canine, feline, equine, bovine, and avian. I am deeply indebted to the 10 contributing authors, whose veterinary expertise in particular animals and medical conditions has enhanced the breadth of this book.

I would also like to thank the following readers for their feedback on the cases as they were being developed: Dharani Ajithdoss, Washington State University; Katie Anderson, University of Minnesota; Gayle Brown, Iowa State University; Chris Chase, South Dakota State University; Jim Collins, University of Minnesota; Philip H. Elzer, Louisiana State University; Keke Fairfax, Purdue University, Catherine E. Hagan, University of Missouri; Stephanie R. Ostrowski, Auburn University; Susan Tornquist, Oregon State University; Dirk Werling, University of London; Elia Tait Wojno, Cornell University; Xiaoping Zhu, University of Maryland; Heather Zimbler-DeLorenzo, Alfred University; Annetta Zintl, University College Dublin.

Katie Laurentiev at Garland Science has played a pivotal role in bringing this project to fruition. I cannot thank her enough for her friendly persistence and helpful editorial advice.

Lastly, I thank my husband, Eric Gershwin, who has provided unending encouragement.

INSTRUCTOR RESOURCES WEBSITE

The images from *Case Studies in Veterinary Immunology* are available on the Instructor Resources website in two convenient formats: PowerPoint® and JPEG. They have been optimized for display on a computer. The resources may be browsed by individual cases, and there is a search engine. Figures are searchable by figure number, by figure name, or by keywords used in the figure legend from the book.

Accessible from www.garlandscience.com, the Instructor Resources website requires registration and access is available only to qualified instructors. To access the Instructor Resource website, please contact your local sales representative or email science@garland.com.

CONTRIBUTORS

Cases 2 and 10 have been contributed by John Angelos, Professor, Department of Medicine & Epidemiology, School of Veterinary Medicine, University of California, Davis.

Case 8 has been contributed by Brian G. Murphy, Associate Professor, Department of Pathology, Microbiology & Immunology, School of Veterinary Medicine, University of California, Davis.

Case 9 has been contributed by Ellen Sparger, Professor, Department of Medicine & Epidemiology, School of Veterinary Medicine, University of California, Davis.

Case 11 has been contributed by Maurice Pitesky, Lecturer and Assistant Specialist in Cooperative Extension, Department of Population Health & Reproduction, School of Veterinary Medicine, University of California, Davis.

Case 12 has been contributed by Rodrigo A. Gallardo, Assistant Professor, Poultry Medicine Program, Department of Population Health and Reproduction, School of Veterinary Medicine, University of California, Davis.

Case 13 has been contributed by Amelia Woolums, Professor, Department of Pathobiology and Population Medicine, College of Veterinary Medicine, Mississippi State University.

Case 14 has been contributed by Carol Reinero, Associate Professor, Department of Veterinary Medicine & Surgery, College of Veterinary Medicine, University of Missouri.

Case 21 has been contributed by Verena Affolter, Professor, Department of Pathology, Microbiology & Immunology, School of Veterinary Medicine, University of California, Davis, and by Catherine Outerbridge, Associate Professor, Department of Medicine & Epidemiology, School of Veterinary Medicine, University of California, Davis.

Case 22 has been contributed by G. Diane Shelton, Professor, Department of Pathology, School of Medicine, University of California, San Diego.

This book is dedicated to the 4431 DVM students to whom I have taught immunology during my tenure on the UCD School of Veterinary Medicine faculty, and to the late John W. Osebold, who introduced me to immunology when I was a second-year veterinary student, and who in subsequent years was my PhD mentor.

CONTENTS

CASE 1
PRIMARY CILIARY DYSKINESIA

The lung is kept free from microorganisms and other inhaled particles by several innate immune mechanisms. These include the turbinate bones in the nose, which are coated with sticky mucosa to entrap large inhaled particles. Particles that are able to move beyond the turbinate bones are removed by a functional epithelium in the bronchi and bronchioles of the respiratory tract, consisting of ciliated epithelial cells and mucus-producing goblet cells. Together, these structures form a defense mechanism (called the mucociliary escalator system) that prevents the movement of inhaled particles and microorganisms into the lungs. Pathogens are trapped in the mucus layer, and the rhythmic beating of the cilia then moves this mucus "blanket" towards the pharynx, where the trapped material is expelled (Figure 1.1). When the cilia are structurally defective, they are not able to create the "wave" of motility necessary for defense, and they thus pose a threat to pulmonary health.

THE CASE OF ANGEL: A YOUNG DOG WITH CHRONIC COUGHING AND REGURGITATION

SIGNALMENT/CASE HISTORY

Angel is a 7-month-old spayed female Old English Sheepdog who was purchased at 3 months of age by her owner, began coughing shortly afterwards, and has continued to do so ever since. She sometimes expels froth during her fits of coughing, and she also vomits and regurgitates during coughing episodes and after meals. She has previously been seen by a veterinarian for dehydration, and has been on and off antibiotic treatment for several months. She is currently on Clavamox 62.5 mg twice a day and Pepcid, and has received all of her core vaccines, but is not on a heartworm preventative at present. During a recent episode of acute respiratory distress that occurred when the owner ran out of antibiotic, the referring veterinarian identified areas of consolidation in Angel's left cranial lung lobe.

PHYSICAL EXAMINATION

On physical examination, Angel was bright and alert (Figure 1.2). Her temperature was 102.6°F (high normal) and her respiratory rate was high at 85 breaths/minute. She did not have any heart murmurs, and her mucous membranes were pink, with a capillary refill time of less than 2 seconds (normal). On auscultation her lung fields were clear, but a cough was elicited by tracheal palpation. Presumably the current antibiotic therapy had resolved the consolidation previously identified in her left lung lobe.

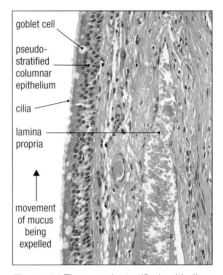

Figure 1.1 The pseudostratified epithelium lines the trachea and consists of ciliated columnar epithelial cells interspersed with goblet cells that contain mucus. The mucus is secreted into the airway lumen, where the cilia create a movement towards the pharynx. The mucus-trapped foreign particles are expelled by the cough reflex. (From Aughey E and Frye FL [2001] Comparative Veterinary Histology with Clinical Correlates, 2nd ed. Courtesy of CRC Press.)

TOPICS BEARING ON THIS CASE:

Innate immunity of the respiratory system

Mucociliary escalator

Inherited defect predisposing to recurrent pneumonia

Figure 1.2 Angel at 7 months of age. (Courtesy of iStock, copyright Sasha Fox Walters.)

DIFFERENTIAL DIAGNOSIS

In the absence of an obvious acute infectious disease, as indicated by lung auscultation and lack of fever, causes of a chronic cough could include allergic bronchitis, cardiac disease (including heartworm infection), and—due to the early age of onset and chronicity of the clinical signs—congenital abnormalities in the respiratory tract, such as ciliary dyskinesis. Familial ciliary dyskinesis has been observed in the Old English Sheepdog. Additional possible causes of vomiting and regurgitation include megaesophagus, neoplasia, and chronic gastritis.

DIAGNOSTIC TESTS AND RESULTS

Thoracic radiographs taken during the most recent acute episode showed alveolar infiltrates, primarily in the cranial lung lobes. No evidence of megaesophagus was found. Abdominal radiographs were unremarkable. A complete blood count (CBC) and chemistry panel showed an elevated white blood cell count of 31,680/μL with a neutrophilia of 26,643/μL. Neutrophils were the predominant cell type. Taken together, these data support a diagnosis of bronchopneumonia. Additional studies to evaluate Angel's mucociliary apparatus were performed based on the history of recurrent bronchopneumonia, which is resolved by antibiotic treatment but recurs once antibiotic therapy is discontinued. These tests included bronchoscopy, bronchoalveolar lavage, and a technetium scan to evaluate ciliary motility. Bronchoscopy and bronchoalveolar lavage revealed a mildly inflamed bronchial mucosa, with a moderate purulent and eosinophilic inflammation. Cultures of bronchoalveolar lavage fluid grew only small numbers of mycoplasma, suggesting that the currently ongoing antibiotic therapy had suppressed bacterial growth. Mucociliary function testing by technetium scanning showed a lack of mucociliary escalator function, consistent with primary ciliary dyskinesia. Biopsies of nasal and tracheal mucosa submitted for electron microscopy showed cilia with an abnormal structure (Figure 1.3).

DIAGNOSIS

Angel was diagnosed with bronchopneumonia due to primary ciliary dyskinesia (PCD). Although electron microscopic evaluation of the ciliary structure is not performed for all reported cases of PCD in dogs, in those cases where it is undertaken it is reported that abnormal dynein arms are most common, as well as abnormal orientation of the central microtubule pair. PCD should be considered in cases of recurrent antibiotic-responsive pneumonia, particularly in a purebred dog.

TREATMENT

There is no procedure for repairing the cilia, and therefore the patient must be treated symptomatically. This involves providing appropriate antibiotics

Figure 1.3 (a) The structure of a normal cilium, which has a central pair of microtubules surrounded by nine doublets. Each doublet has inner and outer dynein arms and radial spokes. The presence of these components in the appropriate structure is required for proper ciliary function. In cases of primary ciliary dyskinesia there is a genetic abnormality that disturbs this normal assembly. (b) An abnormal cilium from a dog with PCD, with displacement of the central microtubule complex and displacement of one of the outer doublets. This is only one example of several ciliary abnormalities that have been described in PCD, including absence of the dynein arms.

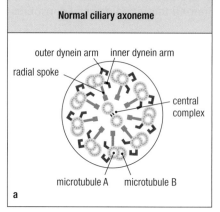
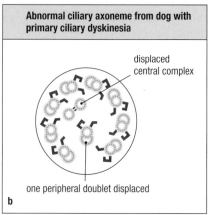

and supportive care. Patients with PCD are encouraged to cough to move the mucus up and out. Percussion can be helpful, and adequate hydration is critical so that the mucus stays fluid. Human patients in rare and extreme cases have needed a lung transplant, but this is not currently an option in dogs.

PRIMARY CILIARY DYSKINESIA

PCD is an inherited condition in which the cilia are not effective. Originally called "immotile cilia syndrome," it is now referred to as primary ciliary dyskinesia because the cilia are rarely completely immotile—most cases have cilia with some degree of motility. However, due to structural defects this motility is aberrant and weak, and is not effective in creating a wave of motility. The case history for Angel is quite typical, with pneumonia that is responsive to antibiotics recurring after cessation of treatment.

A prominent study that reported PCD in a litter of English Pointers reviewed data on PCD presentation in all nine of the puppies. The age of presentation ranged from 7 weeks to 14 months, as expected for a primary inherited defect. Nasal discharge, bronchopneumonia, and leukocytosis were common. A large study on Old English Sheepdogs with PCD identified a genetic mutation affecting the *CCDC39* gene, causing absence of a functional CCDC39 protein that is involved in the motility of cilia (it has an important role in the assembly of the dynein inner arm complex, and is responsible for regulation of the ciliary beating). The dogs in which this recessively inherited mutation was expressed had abnormalities in the central microtubules, and like the English Pointers and Angel they had recurrent nasal discharge, cough, pyrexia, leukocytosis, and bronchopneumonia. The discovery of this genetic link to PCD has provided breeders with the ability to scan the genome for this defect and perhaps ultimately eliminate it from the breed.

COMPARATIVE MEDICINE CONSIDERATIONS

In humans, PCD is a rare disease and is inherited as an autosomal recessive trait. As in dogs, it is characterized by reduced or absent mucus clearance from the lungs. Other organ systems that contain cilia can also show associated dysfunction, including the ear (causing deafness) and the reproductive tract (causing infertility). Renal fibrosis is another commonly associated pathology. Hydrocephalus and abnormal sternebrae are also sometimes associated with PCD in human patients. Some similar abnormalities have been noted in canine patients, including abnormal sternebrae, renal fibrosis, situs inversus, and hydrocephalus.

In human patients a variety of ultrastructural abnormalities have been identified by electron microscopy of the cilia, namely absent or abnormal dynein arms, absence of radial spokes, transposition of microtubules, and random orientation. The majority of human patients with PCD have ultrastructural defects in the proteins that form the cilia and provide their motility. As in dogs, mutations have been identified in genes coding for proteins that form the dynein arms of the cilia. One form of the disease is called Kartagener syndrome, which is defined by the triad of situs inversus, rhinosinusitis, and bronchiectasis. However, the majority of PCD diagnoses, defined by irregular, ineffectual function of the cilia, are not classified as Kartagener syndrome.

Questions

1. This case has focused on the role of the mucociliary apparatus in prevention of infection of the lungs. What other structures or cells are involved in keeping the lungs free from inhaled particles and bacteria?

2. What organs other than the lungs are often affected in a patient with ciliary dyskinesia, and what is the effect of the non-functional cilia on these organs?

3. Angel had been spayed prior to being diagnosed with ciliary dyskinesia. If she had instead been bred to another dog carrying the same mutation, in a litter of four puppies how many individuals would you expect to have the disorder?

Further Reading

Merveille AC, Battaille G, Billen F et al. (2014) Clinical findings and prevalence of the mutation associated with primary ciliary dyskinesia in Old English Sheepdogs. *J Vet Intern Med* 28:771–778.

Morrison WB, Wilsman NJ, Fox LE & Farnum CE (1987) Primary ciliary dyskinesia in the dog. *J Vet Intern Med* 1:67–74.

Wilsman NJ, Morrison WB, Farnum CE & Fox LE (1987) Microtubular protofilaments and subunits of the outer dynein arm in cilia from dogs with primary ciliary dyskinesia. *Am Rev Respir Dis* 135:137–143.

CASE 2
LEUKOCYTE ADHESION DEFICIENCY

Polymorphonuclear leukocytes (neutrophils) are formed in the bone marrow and then enter the bloodstream, where they circulate for a short time. In response to signals from sentinel cells, pathogens, and cytokines, these cells will move out of the blood vessels and into the tissues, where they can act as phagocytes to protect the host from bacterial infection. This movement is dependent upon correct expression of a family of cell-surface glycoproteins called $\beta 2$ integrins. The neutrophils respond to an infectious agent by the binding of leukocyte carbohydrate ligands to selectins on the activated vascular endothelial cells (postcapillary venules and capillaries); they bind transiently to the selectins and begin to roll along the endothelium (Figure 2.1). Under the influence of CXCL8 (IL-8), $\beta 2$ integrins (LFA-1 and MAC-1) are expressed on the leukocyte and bind to the induced adhesion molecule ICAM-1 on the vascular endothelium. Binding of $\beta 2$ integrins to endothelial cells of blood vessels brings the rolling leukocytes to a stop and allows them to begin the process of diapedesis between endothelial cells in response to chemotactic factors and migration to the area of infection (Figure 2.2). In the absence of functional $\beta 2$ integrins the neutrophils are unable to exit the blood vessels. Despite adequate bone-marrow production of neutrophils, these cells are unable to reach the sites of infection where they are most needed. Animals affected by such defects consequently exhibit severe neutrophilia and persistent infections.

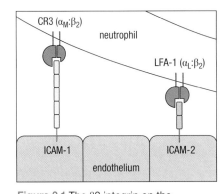

Figure 2.1 The $\beta 2$ integrin on the neutrophil binding to ICAM-1 on the endothelial cell, which occurs in the fluid phase of the acute inflammatory process. It is followed by slowing of local blood flow with margination "pavementing" of the neutrophils along the blood vessel endothelium, and finally the movement of the neutrophils via diapedesis between endothelial cells. (From Murphy K & Weaver C [2017] Janeway's Immunobiology, 9th ed. Garland Science.)

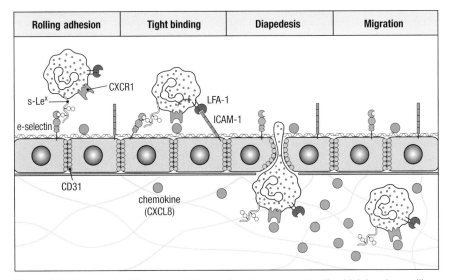

Figure 2.2 Movement of the neutrophil along the blood vessel wall, which involves rolling adhesion, tight binding, diapedesis, and finally migration of the neutrophil between the endothelial cells to travel to the site of infection. (From Geha R & Notarangelo L [2016] Case Studies in Immunology, 7th ed. Garland Science.)

TOPICS BEARING ON THIS CASE:

Migration and homing of leukocytes

Adhesion molecules

Phagocytic cell defects

Genetic causes of immune deficiency

Figure 2.3 Holstein bull calf Tag 5096 showing signs of weakness and ill thrift. (Courtesy of John Angelos.)

THE CASE OF TAG 5096: A CALF WITH UNINHIBITED BACTERIAL INFECTIONS AND GENERAL ILL THRIFT

SIGNALMENT/CASE HISTORY

Tag 5096 is a 10-week-old Holstein bull calf born on a small dairy farm. A couple of months after birth, he developed pneumonia that proved unresponsive to antibiotic therapy, and a general ill thrift and weakness.

PHYSICAL EXAMINATION

On physical examination, Tag 5096 was quiet and alert but unable to stand (Figure 2.3). The calf was febrile (temperature 103.5°F; normal range, 100.5–103°F), tachycardic (heart rate 140 beats/minute; normal range, 60–80 beats/minute), and tachypneic (respiratory rate 60 breaths/minute; normal range, 20–40 breaths/minute). Dried feces were present on the calf's tail, and the animal was noted to have profuse watery diarrhea. Thoracic auscultation revealed bilateral crackles and wheezes that were most pronounced in the cranial ventral areas of the lung fields. Severe scleral injection was also present. Oral examination revealed periodontal gingivitis and gingival recession around the incisors.

DIFFERENTIAL DIAGNOSIS

Recumbency was attributed to the pneumonia and diarrhea, with dehydration and possible electrolyte imbalance. Other potential conditions that could lead to recumbency in calves include nutritional myodegeneration associated with a vitamin E or selenium deficiency, or with deficiency of another trace mineral, such as copper. Because the calf was born on a dairy farm that had an adequate trace mineral supplementation program, these differential diagnoses were considered unlikely. Swollen joints were not observed. Therefore septic arthritis was not considered a likely cause of recumbency. The presence of severe scleral injection suggested possible septicemia. The abnormal lung sounds were suspected to be due to bacterial bronchopneumonia (either primary or secondary to previous viral infection). Bacterial causes that were considered included *Mannheimia haemolytica*, *Pasteurella multocida*, *Mycoplasma* species, and *Histophilus somni*. Viral causes of pneumonia in young calves may include infectious bovine rhinotracheitis (IBR) virus, bovine respiratory syncytial virus (BRSV), parainfluenza (PI3), and bovine viral diarrhea virus (BVDV). Possible causes of the diarrhea in this 10-week-old dairy calf included bacterial, viral, parasitic, and protozoal enteritis. Frequent antibiotic treatment can also cause diarrhea. Based on the overall clinical presentation of this animal, the assessment was that the calf was suffering from pneumonia and diarrhea with possible concurrent septicemia. Failure of passive transfer of maternal immunoglobulins was suspected. An alternative but less likely diagnosis was bovine leukocyte adhesion deficiency.

DIAGNOSTIC TESTS AND RESULTS

The initial diagnostic plan included a complete blood count (CBC), serum chemistry panel, thoracic radiographs, and fecal culture for *Salmonella* and other bacterial agents. The results of the CBC showed an increased hematocrit (39.1%; normal range, 23–33%), leukocytosis (50,200 white blood cells/μL; normal range, 5800–6800/μL) with neutrophilia (39,156/μL; normal range, 2300–6800/μL) and lymphocytosis (10,542/μL; normal range, 1700–5600/μL). The plasma fibrinogen concentration was at the high end of normal (700 mg/dL; normal range, 300–700 mg/dL), and plasma protein levels were elevated (8.8 g/dL; normal range, 6.5–8.5 g/dL). The chemistry panel indicated a low bicarbonate concentration (17 mmol/L; normal range, 23–32 mmol/L) and slightly elevated blood urea nitrogen levels (28 mg/dL; normal range, 7–18 mg/dL); these changes, which were attributed to the diarrhea, resulted in a metabolic acidosis with probable prerenal azotemia. Globulin levels were increased (6.7 g/dL; normal range, 2.9–5.1 g/dL) and albumin levels were decreased (2.2 g/dL; normal range, 3.1–4.3 g/dL). The increased globulin levels indicated that failure of passive transfer was less likely. Thoracic radiographs revealed a pulmonary alveolar pattern suggesting the presence of infiltrates in the cranioventral and caudoventral lung fields, as well as a diffuse bronchointerstitial pattern throughout the lung fields. These changes were consistent with a severe pneumonia.

In Tag 5096 the most concerning hematological abnormality was the marked mature neutrophilia. Although the calf did have multiple possible sources of infection (such as the lungs and gastrointestinal tract), the magnitude of the neutrophilia coupled with the history of a nonresponsive pneumonia, ill thrift, enteritis, and concurrent gingivitis warranted further investigation for possible bovine leukocyte adhesion deficiency (BLAD), also known as bovine granulocytopathy syndrome. Blood was collected to harvest DNA for testing for the mutations associated with the D128G allele that encodes bovine CD18 responsible for the defective β subunit of β2 integrin (CD11/CD18). Aerobic culture and sensitivity of transtracheal wash fluid yielded *Pasteurella multocida* sensitive to oxytetracycline. The fecal culture results were negative for *Salmonella* species. While awaiting the BLAD test results, the calf was given supportive therapy, including antibiotic and fluid therapy.

DIAGNOSIS

The results of the BLAD test revealed that the calf was homozygous for the D128G mutation; as such, the animal was confirmed to have BLAD. Histologic findings in the ileum were also characteristic of BLAD. There was severe mucosal ulceration with bacterial colonization of the mucosal surface. However, no accompanying neutrophilic infiltration was observed. Inflammation at this location consisted of macrophages and plasma cells. A similar finding was present in some sections of lung, where there was necrosis of bronchiolar epithelium accompanied only by macrophage and plasma cell infiltration. However, in other sections of lung a marked infiltration of neutrophils was present. Neutrophils in these sections were most probably a result of a sluggish pulmonary microcirculation with less turbulence, allowing some defective neutrophils to migrate across vessel walls.

TREATMENT

In cattle, BLAD is not treatable. Based on the poor prognosis, the owner elected to have the calf humanely euthanized and submitted for necropsy examination. Through a rigorous testing program for the disease, the Holstein Breed Association has decreased the presence of carriers for this autosomal recessive defect in artificial insemination (AI) sires.

BOVINE LEUKOCYTE ADHESION DEFICIENCY

Calves that are homozygous for the autosomal genetic mutation (D128G) have nonfunctional β2 integrins (CD18) and develop severe infections such as gingivitis, periodontitis, pneumonia, and loss of teeth. They do not survive. The mutation is a single point mutation from adenine to guanine at position 383 of the *CD18* gene. It results in a nonfunctional leukocyte integrin, which prevents leukocyte adhesion to vascular endothelial cells. Thus the calf with BLAD has an increase in circulating neutrophils in peripheral blood due to continuous stimuli that result from bacterial infection, but few or no neutrophils in the tissues where they are needed.

COMPARATIVE MEDICINE CONSIDERATIONS

In human patients, leukocyte adhesion deficiency type 1 (LAD I) is an autosomal recessive disorder caused by mutations in the *ITGB2* gene, encoding the β2 integrin family. The main features of this disease include severe recurrent infections, impaired wound healing, and periodontal disease. Like Tag 5096, affected human children lack functional β2 integrins and develop recurrent infections despite having very high levels of circulating leukocytes. Severe gingivitis is also commonly seen, due to the importance of leukocytes in defense against oral mucosal infection. Unlike calves, human babies afflicted with LAD are treated with drugs to decrease their own T-cell populations, and are then transplanted with maternal or other matched bone marrow cells with the intent of engrafting the infant's bone marrow with competent leukocytes. It is hoped that in veterinary medicine the BLAD syndrome will soon disappear as the mutation is removed from the gene pool, although new mutations could arise. There appear to be multiple novel mutations present in the *ITGB2* gene, each of which has been associated with LAD in humans.

Questions

1. How do you explain the lack of neutrophils in bacterial ileitis despite the high neutrophil count in the blood of this calf?

2. High globulin levels indicate that the antibody responses in this calf were likely to have been normal. How did the leukocyte adhesion deficiency prevent antibodies from facilitating removal of bacteria from the calf?

3. This disease is caused by expression of a homozygous mutation (D128G) in the gene coding for the CD18 molecule. How did the use of artificial insemination facilitate development of BLAD in the Holstein breed of cattle?

Further Reading

Gerardi AS (1996) Bovine leucocyte adhesion deficiency: a review of a modern disease and its implications. *Res Vet Sci* 61:183–186.

Kehrli ME Jr, Ackermann MR, Shuster DE et al. (1992) Bovine leukocyte adhesion deficiency. Beta 2 integrin deficiency in young Holstein cattle. *Am J Pathol* 140:1489–1492.

Nagahata H (2004) Bovine leukocyte adhesion deficiency (BLAD): a review. *J Vet Med Sci* 66:1475–1482.

Parvaneh N, Mamishi S, Rezaei A et al. (2010) Characterization of 11 new cases of leukocyte adhesion deficiency type 1 with seven novel mutations in the *ITGB2* gene. *J Clin Immunol* 30:756–760.

Schmidt S, Moser M & Sperandio M (2013) The molecular basis of leukocyte recruitment and its deficiencies. *Mol Immunol* 55:49–58.

CASE 3
CYCLIC NEUTROPENIA

Neutrophils (also known as polymorphonuclear leukocytes or PMNs) are an important first line of defense against bacterial pathogens. Formed in the bone marrow from myeloid precursors, mature neutrophils are released into the blood on a regular basis. Each cell has a relatively short life span, spending about 12 hours in the circulation and then 2–3 days in the tissues, where it encounters foreign cells and microbes. The neutrophil is responsive to several chemotactic factors that are released from epithelial and other cells after contact with damage-associated molecular patterns (DAMPs) and/or pathogen-associated molecular patterns (PAMPs). Immunological responses that cause fixation of complement induce potent chemotactic factors, such as C3a and C5a, and these serve to create a gradient toward which the neutrophil travels to reach the site where foreign materials have invaded the body. Once it is near a pathogen, the neutrophil phagocytoses and kills the organism, thus ending its own productive life and becoming a part of the purulent exudate that ultimately escapes the body. It may also be ingested by other phagocytic cells, such as macrophages. This short but efficient life of the neutrophil is imperative for the wellbeing of the host.

In Case 2 of this text we examine neutrophils that are incapable of leaving the blood vascular space in response to chemotactic stimuli, and in Case 4 we examine neutrophils that are unable to ingest and kill. Both of these defects are considered primary immunodeficiencies because they are caused by a genetic mutation. Neutrophils in this case (Case 3) do not exhibit these types of defects. Instead the bone marrow stops making these short-lived cells on a regular cycle, creating critical shortages that render the host unprotected for a period of time.

THE CASE OF TINA: A PALE GRAY COLLIE WITH VARIOUS RECURRENT BACTERIAL INFECTIONS

SIGNALMENT/CASE HISTORY

Tina is a 6-month-old female gray collie dog (Figure 3.1). In her short life she has suffered several episodes of illness, including pneumonia, diarrhea, lethargy, and anorexia. In between these episodes she has had periods of relatively normal health. At the time of examination, Tina had a cough and decreased appetite.

TOPICS BEARING ON THIS CASE:

Innate immunity

Neutrophil function

Genetic disease: primary immunodeficiency

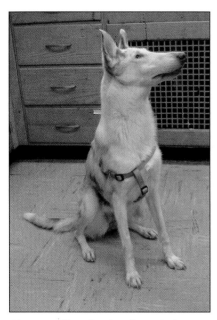

Figure 3.1 Tina showing the typical gray coat coloration associated with the immune defect that causes cyclic neutropenia. (Courtesy of Bill Osborne.)

PHYSICAL EXAMINATION

Tina is somewhat small for her breed and had a body score of 5 out of 10.[1] She was slightly depressed on presentation, and had a moderate seromucoid nasal exudate. On auscultation she had harsh lung sounds bilaterally in the cranioventral region. Upon tracheal palpation a cough could be induced. Her temperature was high at 103.5°F (normal temperature range, 100.5–102°F). All other body systems appeared to be within normal limits.

DIFFERENTIAL DIAGNOSIS

Respiratory disease can be caused by viral or bacterial pathogens. In a young dog, even one with a vaccination history, canine distemper virus should be included as a potential cause of upper and lower respiratory infection. Other respiratory viruses, such as parainfluenza 3 and canine influenza, are also potential causal agents. Bacterial pneumonia, as either a primary or secondary infection, should be investigated. Apart from causes of pneumonia, a thorough differential diagnosis for this case should include potential reasons for frequent recurrent disease. The age of this dog and the frequency of sickness throughout her life indicate that a primary immune defect should be investigated. Primary immunodeficiencies that present with lung infection include selective IgA deficiency, combined immunodeficiency, and neutrophil defects.

DIAGNOSTIC TESTS AND RESULTS

A complete blood count (CBC) and chemistry panel were performed. To evaluate the lung disease, thoracic radiographs and a transtracheal wash were also performed. Immune function was evaluated by a single radial immunodiffusion (SRID) test for IgA levels and enumeration of leukocytes on the CBC. The thoracic radiographs showed increased bronchial density and alveolar infiltrates, indicating that Tina had a bronchopneumonia, which was confirmed as bacterial by the isolation of *Klebsiella pneumoniae* from the transtracheal wash. Polymerase chain reaction (PCR) performed on cells from the transtracheal wash was negative for canine distemper and other respiratory viruses. IgA levels were normal. The CBC revealed a hematocrit of 30 (indicating a slight anemia), and a deficit of neutrophils (946/μL). There was also a hypergammaglobulinemia as indicated by serum electrophoresis.

DIAGNOSIS

Tina was diagnosed with acute bronchopneumonia, secondary to cyclic neutropenia of gray collie dogs (canine cyclic hematopoiesis). The history of recurrent infection and the breed and coat color of the dog strongly suggested cyclic neutropenia. However, to be sure of the diagnosis a follow-up CBC was performed 2 weeks after the initial presentation. At that time the CBC showed a healthy total neutrophil count of 8000/μL. A buccal swab was also submitted to the veterinary genetics laboratory, and the mutation for cyclic hematopoiesis was identified.

TREATMENT

The bacterial pneumonia was effectively treated with amoxicillin. However, the prognosis for the primary defect in neutrophil production was poor. Tina's owners were advised that she would probably have a short life span, and that sterilization was important so that the genetic defect would not be propagated further. The owners were further advised that recent research has shown some positive results from treatment with recombinant granulocyte colony-stimulating factor (G-CSF), using a dose of 3×10^7 IU/kg every 2 weeks for 6 weeks, as well as treatment with a G-CSF in a lentivirus vector. The owners were anxious

---

[1] According to the 2010 AAHA Nutritional Assessment Guidelines for Dogs and Cats.

to do all that they could for Tina, and enrolled her in an experimental therapy program that uses recombinant G-CSF, resulting in normalization of the leukocyte production cycle.

CYCLIC NEUTROPENIA

Cyclic episodes of neutropenia occur in a regular pattern in dogs affected with this condition, leading to episodes of infection when peripheral neutrophil counts fall to less than 1000/μL. The cycle occurs at approximately 10- to 12-day intervals, with periods of neutropenia often followed by a brief neutrophilia (Figure 3.2). Other cell populations also display cycling, but B-cell populations in lymph nodes remain normal. A variety of clinical signs can result, generally related to bacterial infection, including fever, diarrhea, respiratory infection, skin infection, and often joint pain and bleeding episodes.

Canine cyclic neutropenia is caused by a mutation in the *ELA2* gene, which codes for neutrophil elastase. The missense mutation is in the *AP3B1* gene, coding for the β subunit of adaptor protein complex 3. This causes a defect in the G-CSF signal transduction pathway. The mutation is autosomal recessive; thus heterozygotes appear normal and the disease is only expressed in the homozygote with the double recessive mutation.

COMPARATIVE MEDICINE CONSIDERATIONS

Cyclic neutropenia in humans, as in canines, is caused by a mutation in the neutrophil elastase gene (*ELANE*). However, the inheritance in humans is autosomal dominant. The disease is very rare in humans (it affects approximately 1 in 1,000,000 people). Patients present with a variety of infectious diseases that tend to recur with the episodes of neutropenia that occur approximately every 21 days.

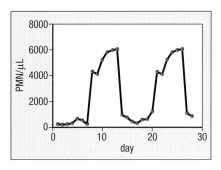

Figure 3.2 Periodicity of peripheral blood neutrophil (PMN) numbers, shown over two cycles. The start of the cycle is defined as the day when the PMN levels fall to less than 1000/μL.

Questions

1. Why does a relatively short period of neutrophil depletion in the periphery have such a disastrous effect on the host?

2. What is the cause of excessive bleeding in some collies that have cyclic neutropenia?

3. When working up a patient with an acute bacterial pneumonia, what difference would be obvious in the test results between the normal dog with pneumonia and the cyclic neutropenia patient?

Further Reading

Dale DC, Rodger E, Cebon J et al. (1995) Long-term treatment of canine cyclic hematopoiesis with recombinant canine stem cell factor. *Blood* 85:74–79.

DiGiacomo RF, Hammond WP, Kunz LL & Cox PA (1983) Clinical and pathologic features of cyclic hematopoiesis in grey collie dogs. *Am J Pathol* 111:224–233.

Horwitz M, Benson KF, Duan Z et al. (2004) Hereditary neutropenia: dogs explain human neutrophil elastase mutations. *Trends Mol Med* 10:163–170.

Pacheco JM, Traulsen A, Antal T & Dingli D (2008) Cyclic neutropenia in mammals. *Am J Hematol* 83:920–921.

Yanay O, Dale DC & Osborne WR (2012) Repeated lentivirus-mediated granulocyte colony-stimulating factor administration to treat canine cyclic neutropenia. *Hum Gene Ther* 23:1136–1143.

CASE 4
CHEDIAK–HIGASHI
SYNDROME

The innate immune system is essential for defense of the host against pathogens during the initial phase of infection, before acquired immune responses are generated. Neutrophils are prominent participants in this early defense against bacterial disease agents. They are produced from stem cells in the bone marrow, and mature neutrophil cells are released into the circulation daily. These cells are called from the blood vasculature to areas of bacterial invasion by chemotactic factors, such as complement components. As described in Case 2, the neutrophils leave the circulation by binding their β2 integrins to adhesion molecules expressed on the endothelial cells that line the blood vessel. Next they undergo a process called diapedesis—that is, a crawling motion that results in the cell moving between the endothelial cells that line the blood vessel, and ultimately exiting the blood vessel into the surrounding tissue. The neutrophils are directed towards the pathogen by creation of a concentration gradient. Chemotactic factors such as C5a, IL-8, and leukotriene B4 are released at the site of pathogen accumulation in the tissue. Once the neutrophil reaches the site of bacterial infection, it attempts to engulf bacteria and kill them. This process, called phagocytosis (Figure 4.1), involves the flowing of cytoplasm around the bacteria, ultimately forming an enclosed vesicle called a phagosome within the cytoplasm of the neutrophil. Most bacteria are easily engulfed and taken into a phagocytic vacuole, but those with heavy polysaccharide capsules may require substances called opsonins to help to bridge the gap between bacteria and the neutrophil membrane. The C3b molecules from complement and antibody are opsonins that can perform this function.

Once they have engulfed bacteria, the lysosomal granules of neutrophils fuse with the phagosome to create a phagolysosome—an environment in which the bacteria are killed. The neutrophil has three types of lysosomal granules, classified on the basis of their content, and each plays an important role in killing engulfed pathogens. The primary or azurophilic granules contain myeloperoxidase, defensins, serum proteases, cathepsin G, and neutrophil elastase. The secondary or specific granules contain lactoferrin, lysozyme, NADPH oxidase, and collagenase. The tertiary granules contain gelatinase and cathepsin. NADPH oxidase is responsible for a respiratory burst, which creates hypochlorous acid, a potent bactericidal agent. When lysosomal granules of any of the three types are larger than normal and/or abnormal in function, this can interfere with the process of phagocytosis.

THE CASE OF SMOKEY: A CAT WITH LIGHT SENSITIVITY AND PERSISTENT BACTERIAL INFECTIONS

TOPICS BEARING ON THIS CASE:

Primary (inherited) immunodeficiency

Neutrophil structure and function

Bactericidal activity

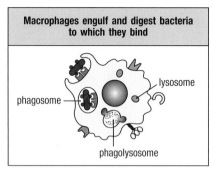

Macrophages engulf and digest bacteria to which they bind

Figure 4.1 The process of phagocytosis involves adhesion, engulfment, formation of a phagosome, and then fusion of lysosomes with the phagosome and killing by oxidative and non-oxidative mechanisms, followed by elimination of bacterial remnants. The engulfment phase involves initial binding of receptors on the phagocyte membrane to bacteria that are to be engulfed. These may be mannose receptors or integrins, but can also be CD32, an Fc receptor that binds antibody Fc which is serving as an opsonin for the bacteria. C3b receptors (CR1) will also bind bacteria that have C3b fixed to the bacterial surface. Once the receptors have bound, there is polymerization of F actin, and the actin-myosin forms a filamentous network of lamellipodia that engulfs the bacterium and draws it into the cell, where it is contained in a vacuole called a phagosome. The destruction of the bacterium is the goal of this process, and it is accomplished by the fusion of the lysosomal granules with the phagosome. With this fusion the cell membrane-associated NADPH oxidase is activated to create toxic oxygen radicals (H_2O_2 and $OC1^-$), which damage and kill the engulfed bacteria. The myeloperoxidase catalyzes the conversion of H_2O_2 into toxic radicals. Other granule contents have additional bactericidal effects: lysozyme breaks up bacterial cell walls, elastase and cathepsin G degrade connective tissue, and antimicrobial peptides exert a bactericidal effect by creating pores in bacterial cell membranes and causing membrane disruption. (From Geha R & Notarangelo L [2016] Case Studies in Immunology, 7th ed. Garland Science.)

SIGNALMENT/CASE HISTORY

Smokey, a 4-month-old male cat, was born into a litter of prize-winning inbred blue smoke Persian cats. The breeder noticed that Smokey and one other littermate appeared to be very sensitive to light (that is, they exhibited photophobia), and they also both had a light yellow/green eye color. During his first few months of life, Smokey developed a bacterial conjunctivitis and appeared to show signs of a respiratory infection on two separate occasions. When he was presented to the veterinarian for castration, a pre-surgical blood sample was taken. It was noted that the site of venipuncture took longer than expected to stop bleeding; this led to further examination.

PHYSICAL EXAMINATION

Smokey was noted to have a gray hair coat and yellow/green eye color (Figure 4.2). On ophthalmic examination he had decreased pigmentation of the tapetum. He also showed marked photophobia when a light was shone into his eyes. He was alert and responsive. Abdominal palpation revealed a slight enlargement of both the liver and the spleen; peripheral lymph nodes were within normal limits. Smokey's lung fields auscultated with a slightly harsh quality bilaterally, particularly in the anterior-ventral lung fields. His mucous membranes were pink, and capillary refill was within normal limits. Moderately severe periodontal disease was present, which was particularly notable for such a young cat. When blood was taken for further diagnostic testing, a prolonged bleeding time was again observed.

DIFFERENTIAL DIAGNOSIS

The young age of onset for frequent infection makes a primary immunodeficiency a likely diagnosis. This could be a neutrophil or antibody defect, because the infecting agents are bacteria. IgA or IgG deficiency would have to be manifested by this age because maternal immunity would have waned before 4 months. The prolonged bleeding time could indicate a clotting defect, possibly thrombocytopenia or a deficiency of one or more clotting factors. One syndrome might explain both defective neutrophil function and impaired platelet function, namely Chediak–Higashi syndrome.

DIAGNOSTIC TESTS AND RESULTS

A complete blood count (CBC) showed that Smokey's neutrophils had an abnormal morphology. Each contained one or more abnormally large structures that were identified as lysosomes (Figure 4.3). Platelet count and neutrophil numbers were also low. Erythrocyte parameters were normal. Quantitative IgG and IgA concentrations were determined by single radial immunodiffusion (SRID), and were found to be within the normal range for cats.

Evaluation of neutrophil function involves several important measures, namely number (to determine whether there are sufficient neutrophils being produced), the ability to respond to chemotactic stimuli (tested by chemotaxis assay), and the ability to engulf bacteria (evaluated using a bactericidal test). Testing for the latter two functions was not undertaken in Smokey's case because the abnormally large granules, together with the gray hair coat, are highly indicative of Chediak–Higashi syndrome. If those assays had been performed, they would probably have shown depression of both chemotaxis and bactericidal activity.

DIAGNOSIS

Based on the abnormal neutrophil granules, light pigmentation, breed of cat, increased bleeding, and frequent bacterial infections, Smokey was diagnosed with Chediak–Higashi syndrome.

TREATMENT

Treatment for Chediak–Higashi syndrome is symptomatic. Smokey's owner was informed that he should not be bred, and that his sire and dam were carriers of the gene responsible for this abnormality. It was recommended that neither of those cats should be bred again. As Smokey developed new infections, each was treated with antibiotics. The owner was also informed that Smokey and his affected sibling were at high risk for cataract development, and that these cats often have a much shortened life span. Bleeding is common, so avoidance of trauma was also advised. Allogeneic bone-marrow transplants are reported to have corrected the neutrophil migration defect and the platelet defect in some patients, but this treatment option was not pursued in this case.

CHEDIAK–HIGASHI SYNDROME

Chediak–Higashi syndrome is an inherited defect observed in Persian cats with the blue smoke coat color. The product of an autosomal recessive mutation, it results in neutrophil lysosomes that are often both larger than normal, and abnormal in function. The abnormal granules prevent normal neutrophil killing of engulfed organisms (phagosomes and lysosomes are unable to fuse because the lysosomes are defective), and neutrophils are less agile in chemotaxis. This impairment of neutrophil activity often leads to infection. Granules in other cells, including basophils, eosinophils, and natural killer (NK) cells, are also enlarged. Abnormally large lysosomes in platelets can result in a prolonged bleeding time. The syndrome is associated with the diluted blue smoke coat color because of abnormalities in melanin granules in melanocytes, which can be observed in the hair shafts. Cats with this syndrome also usually have a yellow-green eye color, and are frequently very sensitive to light. Formation of cataracts in the eyes, which is common, is thought to be caused by the release of peroxidases from the granules, which rupture more easily because of their abnormal structure.

COMPARATIVE MEDICINE CONSIDERATIONS

Chediak–Higashi syndrome is a rare autosomal recessive disorder that occurs in humans and has been described in a variety of animal species, including Aleutian mink, orca whale, cattle (Hereford, Brangus, and Japanese Black), and beige mice. In affected species the mutation is in the *CHS1/LYSY* gene, which is associated with the production of the lysosomal trafficking regulator protein. Mutations have been identified throughout this gene, and some induce a more severe disease than others. The clinical signs include increased incidence of infections, mild coagulation defects, and neurological problems. Like affected cats, human patients have abnormally large lysosomes, and neutrophil numbers are normally low in the periphery. Because of abnormal function of cytotoxic T cells and NK cells (they are unable to release granules for killing), viral infections are common. Individuals with the syndrome also have light, silvery hair (oculocutaneous albinism). In affected children, infection with Epstein–Barr virus causes an accelerated syndrome which presents like lymphoma, with rapid division of leukocytes throughout the body.

Chediak–Higashi syndrome occurs in a mouse strain called CHS1/LYST/Beige. The beige mouse therefore provides a good model for studies of this disease.

Questions

1. Patients with Chediak–Higashi syndrome have abnormal neutrophil granules. Compare the immune defects in this case with those seen in a calf

Figure 4.2 Smokey showing the gray hair coat color and yellow/green eye color typical of blue smoke Persians with Chediak–Higashi syndrome. (From Moriello KA & Diesel A [2014] Small Animal Dermatology. Volume 2: Advanced Cases. Courtesy of CRC Press.)

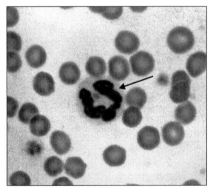

Figure 4.3 Blood smear showing a neutrophil from Smokey's CBC. The arrow indicates large lysosomal granules (eosinophilic) in the cytoplasm of the neutrophil. The enlarged granules seen in Chediak–Higashi syndrome neutrophils are a result of fusion of primary and secondary granules. These cells are poorly chemotactic and have a significantly decreased ability to kill bacteria. The enlarged lysosomes are also present in cells that carry pigment (for instance, melanocytes), which results in a dilution of the pigment in the hair and eyes. (From Moriello KA & Diesel A [2014] Small Animal Dermatology. Volume 2: Advanced Cases. Courtesy of CRC Press.)

with BLAD (Case 2) and in a gray collie with cyclic neutropenia (Case 3). In what respects are they similar and how do they differ?

2. Chediak–Higashi syndrome occurs in beige mice as well as in cats, humans, and several other species. The link between the disorder and gray coat color in cats is explained by abnormal melanin granules in hair shafts and in neutrophils. How is pigment abnormality linked to the disorder in the beige mouse?

3. Why would bone-marrow transplantation be likely to improve the immune function of a patient with Chediak–Higashi syndrome?

Further Reading

Dinauer MC (2014) Disorders of neutrophil function: an overview. *Methods Mol Biol* 1124:501–515.

Kaplan J, De Domenico I & Ward DM (2008) Chediak–Higashi syndrome. *Curr Opin Hematol* 15:22–29.

Kramer JW, Davis WC & Prieur D (1977) The Chediak-Higashi syndrome of cats. *J Lab Invest* 36:554–562.

Shiraishi M, Ogawa H, Ikeda M et al. (2002) Platelet dysfunction in Chediak-Higashi syndrome-affected cattle. *J Vet Med Sci* 64:751–760.

Ward DM, Shiflett SL & Kaplan J (2002) Chediak-Higashi syndrome: a clinical and molecular view of a rare lysosomal storage disorder. *Curr Mol Med* 2:469–477.

CASE 5
SEVERE COMBINED IMMUNODEFICIENCY

EQUINE

The acquired immune response to pathogens depends on functional T and B lymphocytes in the primary and secondary lymphoid organs. During development, the bone marrow nurtures lymphocytes to become B lymphocytes (in avian species this process occurs in the bursa of Fabricius). These B cells ultimately populate the lymph nodes, tonsils, Peyer's patches, and spleen. T lymphocytes develop in the thymus, where they are selected for lack of strong reactivity with self-antigens, and migrate to the same secondary lymph organs as the B lymphocytes. The healthy neonate will have a full complement of T and B cells in secondary lymphoid organs and circulating lymphocytes.

The process of maturation of both cell types involves the synthesis of receptor molecules that are critical for future recognition of antigens. These receptor molecules are synthesized from segments of DNA. To form a T- or B-cell receptor, variable region genes, diversity genes, and joining genes come together to form unique receptors for each cell. Several enzymes control this process, which involves cutting and splicing of the DNA. T- and B-lymphocyte receptors are members of the immunoglobulin superfamily and share similarities—more specifically, they both contain variable and constant regions (Figure 5.1). Both types of lymphocytes recognize antigens through the variable regions of their receptors. In the germline configuration for receptor composition there are multiple potential gene segments at the variable and joining loci. Figure 5.2 demonstrates how the selected sequences are brought into close proximity to form the receptor. Rearrangement of these genes requires excision of the piece of DNA that is not needed, followed by rejoining of the ends to form the rearranged gene. This process occurs for both T- and B-cell receptors, although the genes and segments are different for each. The enzymes responsible for this process include a DNA-dependent protein kinase, which repairs the break in the DNA strand and allows the new receptor to be created from an intact DNA strand. This DNA-dependent protein kinase is a large multi-component enzyme. Defects in one or more of its components can lead to a nonfunctional enzyme, resulting in an inability to make T- and B-cell receptors, and thus an absence of functional T and B cells. Animals without T and B cells in the lymphoid organs are unable to mount an acquired immune response to pathogens.

THE CASE OF SHALIMAR: A FILLY WHO WAS HEALTHY AND FRISKY FOR HER FIRST 5 WEEKS OF LIFE AND THEN DEVELOPED A RESPIRATORY INFECTION

TOPICS BEARING ON THIS CASE:

T- and B-cell antigen-binding receptor formation

Gene rearrangement

Primary (inherited) immunodeficiency

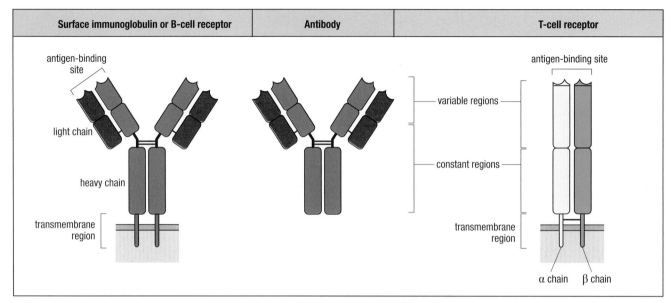

Figure 5.1 The B-cell receptor has variable and constant regions on both light chains (red) and heavy chains (blue); it is essentially an immunoglobulin molecule with attachment to the cell membrane. The variable region on the B-cell receptor provides specificity for antigen binding. T-cell receptors consist of an α chain (yellow) and a β chain (green). Both have a variable region and a constant region. Together the α and β variable regions form a specific binding site for an antigenic peptide held in the groove of a major histocompatibility molecule on the membrane of an antigen-presenting cell. (From Parham P [2009] The Immune System, 3rd ed. Garland Science.)

SIGNALMENT/CASE HISTORY

Shalimar is a chestnut Arabian filly born on a breeding farm for Arabian horses (Figure 5.3). She suckled vigorously after birth, and a SNAP® test for IgG performed at 2 days of age showed that she had obtained adequate amounts of IgG (> 800 mg/dL) from her dam's colostrum. She thrived for her first 5 weeks of life, and then at about 6 weeks of age began to develop a nasal discharge, cough, and conjunctivitis.

PHYSICAL EXAMINATION

Shalimar's body condition appeared to be good, but she was somewhat depressed. Auscultation of her lung fields revealed the presence of harsh lung sounds. She had a mucopurulent bilateral nasal discharge and bilateral conjunctivitis. No peripheral lymph nodes could be located by palpation.

DIFFERENTIAL DIAGNOSIS

The cause of Shalimar's lung disease could be bacterial (for example, *Rhodococcus equi*) or viral (for example, influenza or adenovirus). In an Arabian foal of this age with adequate passive immunity, severe combined immunodeficiency (SCID) as the predisposing cause of lung disease must be considered. Equine adenovirus is a common infection in foals with SCID. *Pneumocystis carinii* is another opportunistic pathogen that is sometimes seen in cases of SCID.

DIAGNOSTIC TESTS AND RESULTS

A complete blood count (CBC) was performed. The most notable aspect of the CBC was an almost complete absence of lymphocytes (Table 5.1). Red blood cell (RBC) parameters and platelet numbers were within normal limits. Fibrinogen levels were high at 800 mg/dL (normal range, 100–300 mg/dL). Equine adenovirus 1 was isolated in tissue culture using a pharyngeal swab and a conjunctival swab; the presence of virus was confirmed by polymerase chain reaction (PCR). Intradermal testing with mitogen was performed using the T-cell mitogen phytohemagglutinin (PHA) injected intradermally into the shaved skin of the neck. After 72 hours the site was examined for the presence of a raised erythematous lesion. Such a lesion would be expected after PHA

Figure 5.2 The process by which variable and joining genes are selected, and those not selected are "looped out" from the DNA and excised. The rearranged gene is used to code for the T-cell receptor (a similar process occurs in the B-cell receptor, but a diversity gene is included). The DNA-dependent protein kinase provides the "paste" to create a rearranged gene. (From Parham P [2009] The Immune System, 3rd ed. Garland Science.)

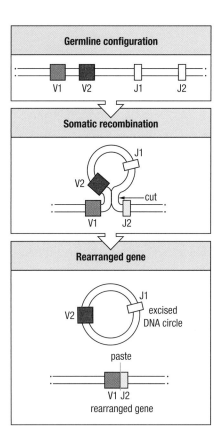

injection into a normal horse because PHA is a T-cell mitogen and will elicit a delayed-type hypersensitivity response. In this case no lesion was present. Quantitative serum IgM and IgG levels were determined. IgM was not detectable, and IgG was detected at 325 mg/dL (normal values are higher than 1000 mg/dL).

DIAGNOSIS

The absence of lymphocytes in the peripheral blood, the lack of IgM in a foal of this age, and the lack of response to injection of a T-cell mitogen strongly supported the diagnosis of severe combined immunodeficiency (SCID).

TREATMENT

There is no available treatment for SCID; it is uniformly fatal in horses. Shalimar was treated symptomatically for her respiratory infection, and when it progressed further and it was clear that the diagnosis was SCID, she was humanely euthanized. At necropsy her lungs showed diffuse consolidation, indicating interstitial pneumonia, and equine adenovirus 1 was isolated. There was no functional lymphoid tissue in the thymus (which consisted of epithelial cell remnants) or in the spleen, and no lymph node structures were present.

SEVERE COMBINED IMMUNODEFICIENCY

SCID in the Arabian horse breed is caused by a genetic mutation in the gene that codes for the p350 subunit of DNA-dependent protein kinase, which is the enzyme responsible for ligating the break formed in the DNA after V-D-J recombination to form the T (V-J) or B (V-D-J) lymphocyte antigen-binding receptors. Without the ability to fix the break in the DNA, no T- or B-cell receptors are formed; the lymphocytes do not survive, and they fail to populate the peripheral lymphoid organs. Thus the horse is left without T and B lymphocytes, so essentially it lacks an adaptive immune response. Such a foal will appear normal while there is coverage from maternally derived immunity. However, when the antibody levels begin to wane at 6–8 weeks of age, the animal is left with no defense, and pathogens that are normally handled efficiently by the immune system (such as equine adenovirus) cause severe disease and ultimately death.

The disease is inherited as an autosomal recessive trait. Thus a horse that is heterozygous for the gene is a healthy carrier, but if two such carriers are mated, the homozygous progeny will have SCID. The defect has been characterized as

Table 5.1 Differential counts for leukocytes in peripheral blood

Parameter evaluated	Patient's result (cells/μL)	Normal range (cells/μL)
Total white blood cells	32,100	5300–14,000
Neutrophils	31,137	3400–11,900
Lymphocytes	0	700–2900
Monocytes	963	50–500
Eosinophils	0	0–100
Basophils	0	0–100

Figure 5.3 Shalimar at 5 weeks of age, immediately before she developed signs of respiratory disease. (Courtesy of iStock; copyright acceptfoto.)

a loss of five nucleotides that causes a frame-shift mutation, resulting in premature termination of the peptide chain with a loss of the kinase domain. Premortem diagnosis of SCID is based on the absence of or very few lymphocytes in peripheral blood, no evidence of antibody production in the foal (no IgM; varying amounts of IgG are usually present from colostral transfer), and the lack of responsiveness to a T-cell mitogen. The absence of functional lymphoid tissue in the thymus and all secondary lymphoid organs is also diagnostic of SCID. There is now a PCR test available to evaluate horses for this mutation, thus helping to prevent the breeding of SCID foals.

COMPARATIVE MEDICINE CONSIDERATIONS

SCID occurs in humans. A variety of syndromes have been described, each caused by a different mutation. The most common of these is the X-linked severe combined immunodeficiency. It is caused by a mutation in the common γ chain that is part of several important immune-cell receptors, namely IL-2, IL-4, IL-7, IL-9, IL-15, and IL-21. Since these molecules are needed for normal immune cell functioning, there is essentially no T- and B-cell function. Untreated babies usually die within a year, due to infection with opportunistic and other pathogens. Unlike foals with SCID, which are usually euthanized or die anyway when diagnosed with this disease, human patients were in the past kept alive in a sterile environment—a so-called "bubble"—to prevent infection. Since those early days of diagnosis, treatment has improved to include bone-marrow transplantation and gene therapy. Other forms of human SCID occur as a result of different mutations. For example, a deficiency of adenosine deaminase caused by a mutation results in an immune system in which lymphocytes cannot proliferate. The human syndrome that is most similar to SCID in horses is called the Omenn syndrome, in which mutations in the recombinase activating genes, *RAG-1* and *RAG-2*, cause inactive enzymes and a lack of the V-D-J recombination required for production of functional T and B lymphocytes.

In dogs there are three breeds in which mutations have been described with resultant SCID. The one most similar to equine SCID involves a mutation that causes early termination of the DNA-dependent protein kinase gene in Jack Russell Terriers. The clinical and pathological features are similar to those seen in Arabian foals with SCID. In the Basset Hound and Welsh Corgi breeds, an X-linked mutation in the interleukin 2Rγ gene has been described. This condition is most similar to the X-linked immunodeficiency seen in humans. Affected dogs have severely depressed immune responsiveness, with reduced numbers of T cells and thymic hypoplasia, but normal numbers of B cells. This represents an interesting distinction from the defect that is seen in SCID in horses and humans. SCID occurs in both male and female horses, whereas the X-linked immunodeficiency occurs only in males. Furthermore, in patients with the X-linked disease there is some IgM production, but SCID foals do not produce any detectable levels of IgM.

SCID mice are a genetically manipulated mouse line in which there is a mutation that affects the DNA-dependent protein kinase, resulting in a lack of ability to make T- and B-cell receptors (like the equine SCID). These mice are used extensively in immunological and infectious disease research.

Questions

1. How do you explain the absence of IgM and the presence of IgG in serum from a foal with SCID?

2. Why is the DNA-dependent protein kinase so important for proper immune system function?

3. If two carrier horses (heterozygotes) are bred, what is the likelihood that the foal will express the SCID defect? Will there be any difference in the likelihood of disease expression in the female versus the male progeny? What is the reason for this difference or lack of difference?

4. Is it possible to eliminate SCID from the Arabian horse breed using genetic testing? If so, what technique would be used?

Further Reading

Bell TG, Butler KL, Sill HB et al. (2002) Autosomal recessive severe combined immunodeficiency of Jack Russell terriers. *J Vet Diagn Invest* 14:194–204.

Kuo CY & Kohn DB (2016) Gene therapy for the treatment of primary immune deficiencies. *Curr Allergy Asthma Rep* 16:39.

McGuire TC & Poppie MJ (1973) Hypogammaglobulinemia and thymic hypoplasia in horses: a primary combined immunodeficiency disorder. *Infect Immun* 8:272–277.

Perryman LE (2004) Molecular pathology of severe combined immunodeficiency in mice, horses, and dogs. *Vet Pathol* 41:95–100.

CASE 6
SELECTIVE IgA
DEFICIENCY

IgA is produced at mucosal surfaces and is the predominant antibody made in response to bacterial colonization of the intestine. It serves as a first-line barrier for the intestinal mucosa, providing protection against pathogens, toxins, and allergens. IgA-producing plasma cells are numerous in the intestine, and secretory IgA is important for preventing bacterial overgrowth in the intestinal tract. The presence of secretory IgA in the bronchial lumen effectively protects the respiratory tract from bacterial and viral infections. IgA is also the antibody isotype that is present in secretions as a dimer and in blood as a monomer.

Plasma cells that secrete IgA are found primarily in the lamina propria of mucosal surfaces, such as the bronchial and intestinal epithelium. Transport of the dimeric IgA into the lumen of the gut or bronchus depends on its binding to a polymeric immunoglobulin receptor (pIgR) that is present on the overlying epithelial cells (Figure 6.1). Once the receptor has transported the IgA into the gut lumen, the receptor undergoes proteolytic cleavage and leaves the small extracellular fragment (called the "J chain") attached to the dimeric IgA. This J chain, referred to as secretory component, helps the IgA to bind to mucins in the secretions, and in the intestinal lumen protects against proteolytic digestion by enzymes. Normal levels of IgA are in the range of 20–150 mg/dL in canine serum, 17–125 mg/dL in saliva, and 80–540 mg/dL in fecal extract. There is also age-dependent variation in these levels of IgA, with lower levels being observed in puppies less than 6 months of age compared with adult dogs. A severe deficiency in IgA production can lead to a variety of mucosal system infections.

TOPICS BEARING ON THIS CASE:

Mucosal immunity

Immunoglobulin deficiency

Genetic causes of immunodeficiency

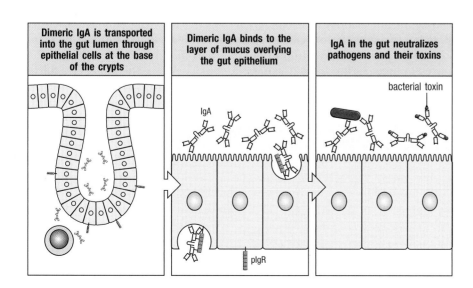

| Dimeric IgA is transported into the gut lumen through epithelial cells at the base of the crypts | Dimeric IgA binds to the layer of mucus overlying the gut epithelium | IgA in the gut neutralizes pathogens and their toxins |

Figure 6.1 IgA is secreted into the intestinal lumen and neutralization of pathogens occurs in the gut as shown. Secretory IgA is also important in the respiratory secretions, where it can neutralize pathogens that enter through the nasal cavity. (From Murphy K & Weaver C [2017] Janeway's Immunobiology, 9th ed. Garland Science.)

Figure 6.2 Dutchess, a young German Shepherd dog with a history of recurrent bouts of bacterial pneumonia. (Courtesy of Pam Eisele.)

THE CASE OF DUTCHESS: A PUPPY WHO HAS HAD A RECURRENT COUGH AND FEVER FOR MOST OF HER LIFE

SIGNALMENT/CASE HISTORY

Dutchess is a 1-year-old spayed female German Shepherd (Figure 6.2) with a history of recurrent bouts of bacterial pneumonia. Since the age of 2 months she has had a moist cough. She has also failed to gain weight as expected for her age and breed. She was first seen by the veterinarian at 4 months of age, presenting with severe cough and a fever. At that time radiographs were taken and she was diagnosed with pneumonia. Treatment with antibiotics for 2 weeks resulted in improvement both clinically and on the lung radiographs. However, at 8 and 11 months, respectively, Dutchess developed pneumonia again and was similarly treated, with good results. Her failure to gain weight has been a continual problem. There is no history of vomiting, but intermittent diarrhea has been reported. She was referred to a specialty practice for further evaluation.

PHYSICAL EXAMINATION

On presentation, Dutchess was thin, with a body score of 3 out of 9,[1] but bright, alert, and responsive. She did not have signs of respiratory disease at the time of her presentation. She did have a strong odor and excessive exudate in the auditory canals. Upon further questioning, the owner confirmed that the dog frequently scratched at her ears.

DIFFERENTIAL DIAGNOSIS

The early onset of recurrent bacterial infection in the respiratory tract, the failure to gain weight, occasional diarrhea, the apparent otitis externa, and the dog's breed suggested that an immunoglobulin deficiency might be involved, particularly IgA, since it is so important on mucosal surfaces. Other possible causes of recurrent pneumonia include aspiration pneumonia due to megaesophagus, ciliary dyskinesia, neutrophil defects, and other anatomic defects. The failure to gain weight could have a variety of causes, including pancreatic insufficiency, food allergy, parasitic infestation, and inflammatory bowel disease.

DIAGNOSTIC TESTS AND RESULTS

Differential diagnoses were ruled out by thoracic and abdominal radiographs (no significant lesions), an esophagram (negative for megaesophagus), a fecal flotation (negative for parasite ova), and a complete blood count (CBC) and chemistry panel. The neutrophil number and morphology were within the normal range. Serum trypsin-like immunoreactivity, cobalamine, and folate levels were also in the normal range. To rule out immunoglobulin deficiency, quantitative IgM, IgG, and IgA levels were determined by single radial immunodiffusion (SRID). The serum IgG level was elevated at 32 mg/mL (normal range, 10–20 mg/mL), the IgM level was slightly elevated at 2.5 mg/mL (normal range, 1–2 mg/mL), and the IgA level was undetectable. Further diagnostics were performed to determine whether secretory IgA was similarly affected; to this end a bronchoscopy and lung lavage were performed. Despite sample concentration by 100-fold, no IgA was detected in the bronchoalveolar lavage fluid (normal control dog lavage, similarly concentrated, had a mean secretory IgA concentration of 1.8 mg/mL).

---

[1] According to the Purina body condition scoring system.

DIAGNOSIS

Dutchess was diagnosed with selective IgA deficiency based on the ruling out of neutrophil defects, parasitism, defects in other immunoglobulin isotypes, and megaesophagus. The undetectable serum IgA and bronchoalveolar lung fluid levels (even after concentration) were pivotal in confirming the diagnosis.

TREATMENT

The inability to make IgA is a genetic defect. Functional IgG, IgM, and cell-mediated immunity are sufficient to prevent the IgA deficiency from becoming life-threatening. Treatment is generally limited to symptomatic treatment of infections with appropriate supportive care as indicated. Thus Dutchess will continue to need antibiotic therapy periodically when she develops respiratory infection and/or infection of other mucosal surfaces.

SELECTIVE IgA DEFICIENCY

Selective IgA deficiency is the most common primary immune deficiency recognized in dogs. It is associated with infections on mucosal surfaces, including the lung, intestinal tract, genital tract, skin, and ear canals. IgA deficiency is also inherited and has been recognized as familial in the following pure-bred dog breeds: German Shepherd, Shar-Pei, Beagle, Airedale Terrier, Basset Hound, Weimaraner, and Irish Setter. The mode of inheritance has not been determined. However, a recent genome-wide association study of four breeds (German Shepherd, Golden Retriever, Labrador Retriever, and Shar-Pei) determined that there are 35 loci associated with low IgA concentrations. The authors of the study suggested that the defect is related to genes that control B-cell development.

The incidence of allergic dermatitis is sometimes increased in dogs that have selective IgA deficiency. It is not unusual to see a compensatory increase in serum levels of the other immunoglobulin isotypes, as was observed in this case for IgG. In German Shepherd dogs an association of aspergillosis infection with low IgA levels has also been observed.

COMPARATIVE MEDICINE CONSIDERATIONS

Primary immunodeficiency in both humans and dogs is characterized by the appearance of infections early in life because the defect is inherited and manifests itself upon loss of maternal antibody protection. Selective IgA deficiency is the most common immunodeficiency disorder in humans, affecting approximately 1 in 700 people of European descent. It can be inherited as either an autosomal dominant or an autosomal recessive trait. Some people are asymptomatic, and others develop a variety of respiratory, gastrointestinal, skin, mouth, and ear infections. Development of autoantibodies to IgA has been reported in some people, while others may spontaneously resolve the defect and begin to produce IgA.

Questions

1. Describe the points during the development of an IgA response when there might be an error that results in selective IgA deficiency.

2. How might the otitis externa seen in Dutchess be related to her IgA deficiency?

3. If Dutchess develops allergies to inhaled allergens, describe how her lack of IgA might contribute to this condition.

4. Do you think that an intranasal vaccine for "kennel cough" (*Bordetella bronchiseptica*/parainfluenza-3) would be effective in a dog with selective IgA deficiency? If not, would you expect a parenteral vaccine to work better?

Further Reading

Driessen G & van der Burg M (2011) Educational paper: primary antibody deficiencies. *Eur J Pediatr* 170:693–702.

Felsburg PJ, Glickman LT, Shofer F et al. (1987) Clinical, immunologic and epidemiologic characteristics of canine selective IgA deficiency. *Adv Exp Med Biol* 216B:1461–1470.

Norris CR & Gershwin LJ (2003) Evaluation of systemic and secretory IgA concentrations and immunohistochemical stains for IgA-containing B cells in mucosal tissues of an Irish setter with selective IgA deficiency. *J Am Anim Hosp Assoc* 39:247–250.

Olsson M, Tengvall K, Frankowiack M et al. (2015) Genome-wide analyses suggest mechanisms involving early B-cell development in canine IgA deficiency. *PLoS One* 10:e0133844.

Pastoret P, Griebwl P, Bazen H & Goaerts A (1998) Immunology of the dog. In Handbook of Vertebrate Immunology, pp. 261–288. Academic Press.

CASE 7
FAILURE OF PASSIVE TRANSFER

The neonate is born into a world filled with pathogens, and the acquired immune response takes some time to develop its full protective potential. For this reason, nature has provided for the very young to obtain immune protection passively from the mother. Depending on the species, this protection may be provided *in utero* (in primates), or after birth in the colostrum. In some species (for example, dogs and cats), both modes of antibody transfer occur. Species that obtain all of their antibody protection from colostrum (for example, ruminants, swine, and horses) must nurse during the first day of life in order to absorb the antibodies provided in the colostrum. Colostral antibodies bind to receptors on the intestinal epithelial cells (in the newborn these receptors are called FcRn), and are endocytosed. After endocytosis they enter lacteals and ultimately the blood, where they circulate to provide passive antibody protection. There are some species differences with regard to which classes of antibody can bind the FcRn. IgG is the main antibody class absorbed to provide protection for the equine neonate. By 24 hours after birth, IgG is no longer absorbed intact from the gut into the blood of the foal. Thereafter, antibodies consumed by the oral route are simply digested as protein, and provide nutrition but no immune protection. This process is called "closure" (Figure 7.1), and it occurs because intestinal cells that bear FcRn are replaced by cells that lack this receptor. A delay in nursing has been shown to slightly extend the window within which absorption can occur. When insufficient levels of antibody have been absorbed before closure occurs, there can be serious repercussions for the health of the young foal.

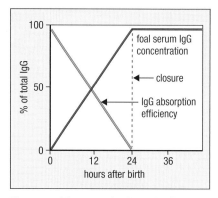

Figure 7.1 The rate of colostral IgG absorption in the neonatal foal that has received colostrum and absorbed immunoglobulins decreases over time (green line). By 24 hours, intestinal closure occurs and immunoglobulin is no longer absorbed intact across the intestinal epithelium into the blood (broken line). The concentration of IgG in the foal's serum (red line) increases until closure occurs.

THE CASE OF DIAMOND: A FOAL WHO DEMONSTRATED SEVERE WEAKNESS AFTER BIRTH

SIGNALMENT/CASE HISTORY

Diamond is a 3-day-old Hanoverian colt (Figure 7.2). He was slightly weak at birth and had to be assisted to stand and nurse. Diamond was the first foal for this mare, and the owner reported that she was anxious and moved around when Diamond attempted to nurse. On his second day of life Diamond's owner thought he might be having trouble breathing, so she called her veterinarian to evaluate him. The veterinarian performed a foal-side SNAP® test, and the result showed that Diamond's IgG level was less than 200 mg/dL. The foal was then taken to the hospital for further evaluation.

TOPICS BEARING ON THIS CASE:

Passive immunoglobulin transfer

Neonatal antibody absorption

Protective levels of IgG for neonate

Figure 7.2 Diamond at 3 days of age. (Courtesy of Shutterstock; copyright Natalia Zhurbina.)

PHYSICAL EXAMINATION

On examination, Diamond's temperature was normal at 102°F, his pulse was 100 beats/minute (foal normal range, 80–100 beats/minute), and his respiratory rate was 60 breaths/minute (foal normal range, 20–40 breaths/minute). He appeared to be normally developed and well muscled, but thin. His mucous membranes were pink and moist, and capillary refill time was normal at 1 second. On auscultation, Diamond's lungs were clear and he did not have any heart murmurs. The foal's joints were not distended. He was bright and alert.

DIFFERENTIAL DIAGNOSIS

The foal's history of having difficulty accessing his dam's colostrum during the critical hours for colostrum absorption makes a provisional diagnosis of failure of passive transfer very likely.

DIAGNOSTIC TESTS AND RESULTS

For rapid determination of whether or not Diamond had received sufficient IgG from colostrum, a SNAP® test was performed. The result showed that his IgG levels were less than 200 mg/dL. This is definitely inadequate, as IgG levels should be at least 800 mg/dL. A quantitative IgG test (SRD) was then performed to determine the actual concentration of IgG in Diamond's blood. It was confirmed as being less than 200 mg/dL. A complete blood count (CBC) and chemistry panel were also performed. Abnormal results included hypoproteinemia at 4.3 g/dL. Hematocrit and white blood cell parameters were within normal limits.

DIAGNOSIS

Diamond was diagnosed with failure of passive transfer based on the history of difficulty nursing and on additional information provided by the SNAP® and SRD test results, which confirmed that the foal had not consumed sufficient colostrum to provide protection.

TREATMENT

An intravenous catheter was placed and Diamond received 2 L of equine plasma. He was also started on 5 mg/kg ceftiofur and 21 mg/kg amikacin prophylactically. The next day he received another 2 L of plasma. His blood IgG levels were tested using a quantitative SRD test, and the results confirmed that he had achieved normal IgG levels (1270 mg/dL). The foal was sent home and his owner was told to watch for increased temperature, joint swelling, cough, and diarrhea. Diamond did not return to the hospital. It was fortunate that he was evaluated and received supplemental IgG before he became infected with bacteria from the environment, which could have resulted in neonatal sepsis—a severe and often fatal condition.

FAILURE OF PASSIVE TRANSFER

Failure of passive antibody transfer affects up to 20% of foals. There are three primary causes: insufficient colostrum from the mare (either in volume or in amount of antibody in the colostrum), inadequate intake by the foal, or failure of absorption through the intestine of the foal. Mares that drip colostrum prior to parturition may decrease their supply for the foal. The IgG content of colostrum may also vary; generally it should be at least 3000 mg/dL to provide sufficient protection for a foal. Vaccination of the equine dam prior to foaling can help to ensure that foals obtain specific protection against common equine pathogens as well as environmental organisms. If a foal is weak at birth or the dam is a poor mother, it should be noted that colostrum intake may not be adequate for protection.

When failure of passive transfer is suspected, there are several methods available for performing a rapid estimate of IgG concentration in a foal. The quantitative "gold standard" is single radial immunodiffusion (SRID), which is based on gel diffusion of the foal's serum antibodies with immune complex formation with antibodies to the IgG. Because the test utilizes known standards and the IgG levels are read off a standard curve, this test is very accurate. However, it is necessary to wait 24 hours before the result can be read, which is generally beyond the window of time in which oral colostral supplementation can still be employed if required. If a foal gets to 24 hours before supplementation, the IgG must be given as plasma by the IV route as in Diamond's case. Other tests can allow the veterinarian to determine whether there is failure of passive transfer early enough to correct the deficit with oral colostrum, even though they are less accurate than SRID. One popular test, the zinc sulfate turbidity test, utilizes the properties of $ZnSO_4$, which combines with IgG to form a white precipitate. The more IgG there is in the sample, the heavier the precipitate. This is an inexpensive and rapid assay for estimating the amount of IgG in a foal's serum. A more recent invention is the SNAP® test, an ELISA-based dot blot format that shows whether a foal has IgG levels of less than 200 mg/dL, 200–400 mg/dL, 400–800 mg/dL, or more than 800 g/dL. A foal with IgG levels of less than 200 mg/dL has failure of passive transfer, and one with levels of 200–400 mg/dL is in partial failure. Foals with IgG levels in the range 400–800 mg/dL may be sufficiently protected if they are kept in a clean environment, but it is preferable to have an IgG level above 800 mg/dL.

COMPARATIVE MEDICINE CONSIDERATIONS

Failure of passive transfer is not a problem in humans and primates, because maternal antibodies are transferred across the placenta during gestation. By contrast, in cattle and other ruminants, failure of passive transfer is the primary cause of neonatal sepsis. It is a common problem in alpacas. In cats and dogs only about 5% of the IgG is transferred through the placenta, and thus consumption of colostrum is also very important.

Questions

1. Explain why it would not be effective to administer colostrum by bottle to a 2-day-old foal with a serum IgG level of less than 200 mg/dL.

2. Why is it a safe procedure not to breastfeed a human infant at all, and instead to feed him or her with formula from birth, whereas this is not an option for a foal or calf? What immunological benefit does the human infant derive from breastfeeding?

3. What might be responsible for a low (200–400 mg/dL) IgG plasma level in a foal that received a sufficient volume of colostrum within the proper time frame?

Further Reading

Crisman MV & Scarratt WK (2008) Immunodeficiency disorders in horses. *Vet Clin North Am Equine Pract* 24:299–310.

Garmendia AE, Palmer GH, DeMartini JC & McGuire TC (1987) Failure of passive immunoglobulin transfer: a major determinant of mortality in newborn alpacas (*Lama pacos*). *Am J Vet Res* 48:1472–1476.

McClure JT, DeLuca JL, Lunn DP & Miller J (2001) Evaluation of IgG concentration and IgG subisotypes in foals with complete or partial failure of passive transfer after administration of intravenous serum or plasma. *Equine Vet J* 33:681–686.

CASE 8
FELINE LEUKEMIA VIRUS

Retroviruses are RNA viruses that, after entering the cell, make a DNA copy (provirus), which is subsequently integrated into the infected cell's genome. Feline leukemia virus (FeLV) is a common retrovirus that infects cats, and has been studied extensively for many decades. It has three genes—*gag, pol*, and *env*—which encode the viral structural proteins (Gag and Env) and the necessary enzymes (protease, reverse transcriptase, and integrase). It also has a host-cell-derived outer lipid membrane, which facilitates inactivation by most disinfectants, including soap. FeLV-infected cats can develop a variety of clinical syndromes, including co-infections, immune suppression, anemia, and tumor formation. Most FeLV-associated tumors are solid and are frequently lymphomas; despite its name, the virus is only rarely associated with leukemia. Virus-induced immunodeficiency often leads to secondary infections (bacterial, viral, protozoal, and fungal). Although the precise mechanisms by which FeLV damages the feline immune system are poorly understood, thymic atrophy, lymphopenia, and neutropenia are common.

Figure 8.1 Sinbad at the time of examination. (Courtesy of Brian G. Murphy.)

THE CASE OF SINBAD: A YOUNG CAT WITH WEAKNESS AND LETHARGY

SIGNALMENT/CASE HISTORY

Sinbad is an approximately 2-year-old spayed female cat (Figure 8.1), who was formerly a stray and was adopted by a family several months ago. Although somewhat skittish, Sinbad is a friendly cat who was spayed prior to adoption. Her vaccination status is unknown. Recently, she has seemed quite lethargic.

PHYSICAL EXAMINATION

On examination, Sinbad's temperature was normal at 100.8°F, and her pulse and heart rate were within normal limits. However, her respiratory rate was high; she breathed through her mouth during the examination (panting). A cough was not easily elicited. No heart murmurs were auscultated, but her respiratory and heart sounds were both subjectively muffled. Her mucous membranes (ocular conjunctiva and oral mucosa) were pale, and her oral capillary refill time (CRT) was moderately prolonged at over 2 seconds. Sinbad's skin, eyes, ears, dentition, and abdominal palpation were all unremarkable. She ambulated without lameness but seemed to be fairly lethargic during the examination. Physical manipulation appeared to be stressful.

TOPICS BEARING ON THIS CASE:

Virus-induced immunosuppression

Immune response to viral infection

Virus-initiated oncogenesis and host response

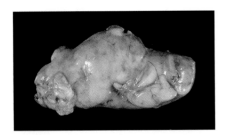

Figure 8.2 The enlarged mediastinal lymph node (7 × 3 × 2 cm) removed from Sinbad at necropsy. (Courtesy of Brian G. Murphy.)

DIFFERENTIAL DIAGNOSIS

Lethargy, pale mucous membranes, and prolonged CRT could be the result of anemia or systemic low blood pressure (shock). The elevated respiratory rate and muffled chest sounds suggest the possibility of a thoracic effusion or mass, which could be either neoplastic or inflammatory.

DIAGNOSTIC TESTS AND RESULTS

A complete blood count (CBC), serum chemistry panel, and two-view chest radiographs were performed. The chemistry panel was unremarkable, and the CBC demonstrated a packed cell volume (PCV) of 19 without a regenerative response (nonregenerative anemia). The radiographs demonstrated a moderate thoracic effusion. Because Sinbad had previously been an unvaccinated stray and was exhibiting lethargy and pale mucous membranes, an in-house feline leukemia virus (FeLV) antigen/feline immunodeficiency virus (FIV) antibody ELISA test (IDEXX) was performed on a whole blood sample. This test showed that Sinbad was positive for FeLV antigen and negative for FIV antibody in blood.

DIAGNOSIS

On the basis of the physical examination and the results of the diagnostic tests, Sinbad was diagnosed with FeLV infection with a nonregenerative anemia and thoracic effusion. Possible causes of the thoracic effusion included pyothorax, chylothorax, neoplasm, and thoracic transudate (cardiogenic effusion). A recommendation for a diagnostic thoracocentesis was declined by the owners, who authorized euthanasia and a subsequent necropsy examination. During the necropsy, approximately 100 mL of blood-tinged fluid were found free in the thoracic cavity. The mediastinal lymph node was greatly enlarged, measuring 7 × 3 × 2 cm (Figure 8.2). The examination was otherwise unremarkable. A complete set of tissue samples was collected, fixed in buffered formalin, and submitted to a diagnostic laboratory for histological examination. The histological findings in the mediastinal lymph node were consistent with a diagnosis of lymphoma (Figure 8.3).

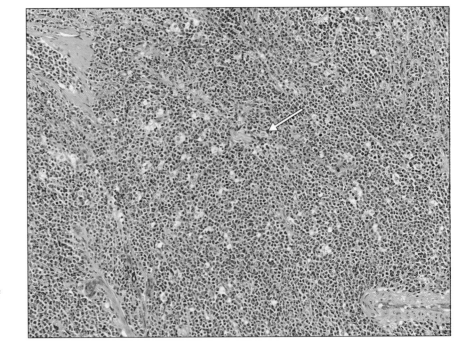

Figure 8.3 Feline lymph node, 200× magnification, hematoxylin and eosin stain. The lymph node architecture is obscured by sheets of neoplastic round cells (lymphocytes), as indicated by the arrow. Lymphadenopathy was apparent grossly and neoplastic lymphocytes showing distinct cell borders, scant cytoplasm, large round nuclei containing heterochromatin, and variable numbers of nucleoli characteristic of lymphoma were seen on histology. (Courtesy of Brian G. Murphy.)

TREATMENT

Once acquired, the feline leukemia virus persists in cats for life, and there is currently no cure for FeLV infection. Treatment is therefore symptom based and supportive. A variety of medical protocols currently exist for leukemia and lymphoma treatment. A killed virus vaccine exists, and use of this vaccine can prevent antigenemia and proviral DNA integration in bone-marrow cells in approximately 90% of young cats if the animals are vaccinated prior to exposure to the virus. Since diagnosis of FeLV depends on detection of virus antigen, vaccination for FeLV does not complicate the interpretation of the FeLV ELISA (the ELISA is targeted at a different viral antigen to the vaccine antigen). By contrast, since the killed virus vaccine for FIV induces antibody formation, use of that vaccine complicates the interpretation of the standard FIV ELISA (antibody-based test). Testing for both FeLV and FIV should be performed before vaccination with those respective vaccines.

FELINE LEUKEMIA VIRUS (FELV)

FeLV is a gammaretrovirus of domestic cats that occurs worldwide. Risk factors for infection include young age, high population density, and poor hygiene. Transmission is often the result of "friendly" contact (mutual grooming) or bite wounds from fighting. Transplacental transmission of FeLV can also occur. As cats age, they generally become more resistant to FeLV infections. Since its discovery the virus has become less common, probably as a result of vaccination, testing, and culling of infected cats.

FeLV infection is associated with cellular transformation and tumor formation. At a molecular level, this can be accomplished in several ways. By chance, proviral integration into the host genome may occur adjacent to a cellular proto-oncogene. The presence of the powerful retroviral promoter adjacent to a proto-oncogene can lead to disrupted gene regulation and uncontrolled cell growth. Cellular transformation can also be initiated if proviral integration directly disrupts a cellular tumor suppressor gene. In addition, viruses that "capture" a host proto-oncogene, such as *v-myc*, and incorporate it into the provirus can become a recombinant acutely transforming virus—for example, feline sarcoma virus (FeSV). In exchange for the cell-derived oncogene, FeSV generally loses a portion of its genome (*gag*, *pol*, or *env*), resulting in a defective virus that requires a helper virus in order to replicate in the infected cells. FeLV-induced lymphomas may be mediastinal/thymic, alimentary, multicentric (affecting peripheral lymph nodes), or atypical (affecting organs, the central nervous system, or skin). As mentioned previously, although the virus is called the feline leukemia virus, leukemia is a less common result of infection. Co-infections with other microbial agents, such as FIV, are relatively common, possibly as a result of immune suppression.

Once infected, cats probably cannot completely clear the FeLV infection. However, some cats fail to develop any clinical signs, are aviremic, and are classified as long-term non-progressors (LTNP). This phenomenon is a result of viral latency in persistently infected cells—there is a lack of viral transcription, and therefore no virus production after proviral integration. Clinical signs generally develop in cats with high viral antigen loads. At the time of writing, the viral or host factors that dictate the type of disease course remain undetermined.

COMPARATIVE MEDICINE CONSIDERATIONS

The feline leukemia virus (a gammaretrovirus) cannot be transmitted to humans. However, there is a human T-cell lymphotropic virus type I (HTLV-I). This is a deltaretrovirus, and it is associated with T-cell leukemia and lymphoma. There is also a closely related family of simian T-cell lymphotropic viruses (STLVs) that infect Old World monkeys. Four types of human T-cell lymphotropic viruses (HTLVs) have been identified, and four types of STLVs are recognized.

Questions

1. Why are FeLV-infected cats likely to be infected for life?

2. What are the possible molecular mechanisms of virus-induced oncogenesis (tumor formation) in cases of FeLV infection?

3. How do the standard diagnostic assays for FeLV and FIV differ? What is the relevance of these differences with regard to vaccination?

Further Reading

Greene CE (ed.) (2006) Infectious Diseases of the Dog and Cat, 3rd ed. Saunders.

Hartmann K (2012) Clinical aspects of feline retroviruses: a review. *Viruses* 4:2684–2710.

Lutz H, Addie D, Belak S et al. (2009) Feline leukaemia. ABCD guidelines on prevention and management. *J Feline Med Surg* 11:565–574.

Stuke K, King V, Southwick K et al. (2014) Efficacy of an inactivated FeLV vaccine compared to a recombinant FeLV vaccine in minimum age cats following virulent FeLV challenge. *Vaccine* 32:2599–2603.

Feline immunodeficiency virus (FIV) is a lentivirus, akin to human immunodeficiency virus (HIV) and simian immunodeficiency viruses (SIVs). These viruses have in common the ability to infect and destroy cells of the immune system, causing acquired immunodeficiency. FIV targets macrophages, CD4$^+$ T cells, CD8$^+$ T cells, B cells, and dendritic cells. It uses two cell surface molecules as receptors: CD134 on activated/memory CD4$^+$ T cells and CXCR4 (α-chemokine receptor-4) on lymphocytes, macrophages, and stem cells. The process by which FIV enters the cell and subsequently replicates is illustrated in Figure 9.1. The ability to use a reverse transcriptase and integrate into the host genome is a characteristic feature of these lentiviruses. Budding of mature virus from the cell surface provides the virions that are present in secretions and blood for propagation of the infection in other host animals.

The acute stage of FIV infection shows a decrease in CD4$^+$ T cells and a high plasma viremia. The number of CD8$^+$ cytotoxic T cells initially rises. Then, during the asymptomatic phase, the CD4$^+$ T-cell concentration in blood will recover, only to subsequently decrease slowly over time. Plasma viremia remains low, and antibody titers initially rise (Figure 9.2). Virus-specific CD8$^+$ cytotoxic T cells eventually decline in number and lose their cytolytic activity due to reduced expression of the effector molecules—granzymes and perforins. Ultimately, during the final phase of the disease (FAIDS), plasma viremia rises and CD4$^+$ T-cell counts in blood decline. The normal ratio of CD4$^+$ to CD8$^+$ T cells decreases (for example, from 1.6 to 0.6). CD4$^+$ CD25$^+$ regulatory T cells producing TGF-β are induced, and they further limit the effectiveness of T-cell effector responses in the infected cat. Dendritic cells are also targets for FIV. There is impaired functioning of dendritic cells, including altered innate immune responses such as cytokine production, and reduced activation of T cells.

THE CASE OF SCOTTY: AN OLDER CAT WITH PERSISTENT INFECTIONS AND A HISTORY OF FIGHTING

SIGNALMENT/CASE HISTORY

Scotty is a 10-year-old indoor/outdoor castrated male domestic short-haired cat who has lived with a family in a rural neighborhood for his whole life. Although he was neutered at 2 years of age, he has frequently been involved in fights with other free-roaming cats living in the same area. Over the course of his life he has been vaccinated for feline leukemia virus, rabies, feline panleucopenia virus, feline herpesvirus, and calicivirus. During the past 6 months, Scotty's weight has gradually declined, although he has not been completely

TOPICS BEARING ON THIS CASE:

Failure of cell-mediated immunity

Infection with feline immunodeficiency virus

Infection with opportunistic pathogens

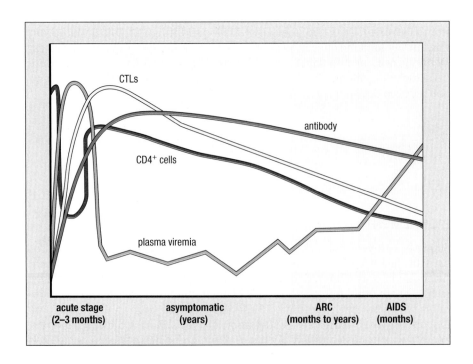

1. Viral attachment to cell surface receptors

viral envelope
viral RNA
capsid protein
reverse transcriptase
CD134
CXCR4

2. Viral fusion with cell membrane

cytoplasm

3. Reverse transcription

4. Nulear translocation and integration into host genome

DNA
nucleus

5. Viral transcription and nuclear export

6. Viral protease and protein processing

7. Virion assembly and maturation

Figure 9.1 Diagram of feline immunodeficiency virus (FIV) replication, with sites for potential antiretroviral therapy indicated: (1) viral attachment to cell surface receptors; (2) viral fusion with cell membrane; (3) reverse transcription; (4) nuclear translocation and integration into host genome; (5) viral transcription and nuclear export; (6) viral protease and protein processing; (7) virion assembly and maturation. (Adapted from Mohammadi H & Bienzl D [2012] *Viruses* 4:708–724.)

anorexic. According to his owner, the cat's appetite has gradually decreased over the same time period, and currently his appetite is poor but he is still eating small amounts of food, and drinking water. The owner has also noted chronic alopecia over Scotty's head and neck, with crusty lesions suggestive of pyoderma, along with chronic sneezing and nasal and ocular mucoid exudate that has gradually worsened over the past 2 months. The owner has not observed pruritus. Scotty is still responsive to his owner, but is depressed. His activity has declined significantly over the past 2 to 3 weeks, and his owner has kept him indoors. His stools are variable, but often soft and sometimes loose, and his stool amount is small because he is eating so much less food.

CTLs
antibody
CD4⁺ cells
plasma viremia

acute stage (2–3 months)   asymptomatic (years)   ARC (months to years)   AIDS (months)

Figure 9.2 The natural history of FIV infection in cats. CTLs = cytotoxic lymphocytes. ARC = Aids Related Complex. (Courtesy of Ellen Sparger.)

PHYSICAL EXAMINATION

Physical examination revealed a severely cachexic, slightly dehydrated, and depressed cat (Figure 9.3), showing alopecia predominantly localized to the head and neck, along with crusty lesions, erythema, and mild pyoderma. Scotty's hair coat in general was scurfy and flaky, indicating that he was not grooming himself. The nares and peri-ocular skin revealed evidence of chronic exudative discharge that was both crusty and mucopurulent. Severe inflammatory proliferative gingivitis was also noted (Figure 9.4). Abnormalities of heart rate, respiration, and body temperature were not observed. Peripheral lymph nodes appeared normal in size. Abdominal palpation did not reveal significant findings or evidence of pain.

DIFFERENTIAL DIAGNOSIS

The history of fighting with outdoor cats and the clinical signs of multiple chronic infections and cachexia suggest infection with FIV. Some of the same clinical signs could be caused by infection with feline leukemia virus (FeLV).

DIAGNOSTIC TESTS AND RESULTS

In order to ascertain Scotty's FIV status, blood was drawn and examined by a quick assay, namely the SNAP® FIV/FeLV test, to determine the presence of antibodies to FIV. The test was positive for FIV and negative for FeLV.

DIAGNOSIS

Based on the test results, Scotty was diagnosed with feline immunodeficiency virus infection. A key to the diagnosis of FIV is symptomatic infection with pathogens that are normally self-limiting in cats, namely upper respiratory viruses, *Toxoplasma*, skin infections (bacterial, fungal, or parasitic), or stomatitis. Infestation with mange mites, infection with *Mycobacterium*, and the presence of nonhealing wounds are also commonly seen.

TREATMENT

There is no commonly used specific treatment for FIV infection, and once the infection is established it is lifelong. However, treatment with the reverse transcriptase inhibitor zidovudine (AZT) at 15 mg/kg administered every 12 hours by the oral or subcutaneous route has been reported to result in some improvement of immune function, and regression of stomatitis. Side effects include anemia. Secondary infections are treated with appropriate antibiotics as they develop, and supportive care is provided as indicated. It is important to alert the cat owner to the potential for disease transmission to other cats. The cat should be confined if possible to prevent dissemination of the virus. Thus in this case Scotty was converted to an indoor-only cat, and received supportive care (antibiotic treatment) for his infections, and fluid therapy to correct the dehydration.

FELINE IMMUNODEFICIENCY VIRUS

The most common mode of infection with FIV in cats is a bite wound from an infected cat. The saliva contains infective virus that is either cell free or cell associated. Once infected, the cat harbors the virus in its genome, and has a slow and persistent course of infection. The virus targets differentiated cells, and has a tropism for lymphocytes and macrophages. Ultimately the cat becomes lymphopenic, with a systemic loss of CD4+ T cells. Loss of helper CD4+ T cells results in opportunistic infections, which can ultimately claim the cat's life.

Clinical and immunological manifestations of FIV infection occur in acute and chronic phases, and have been well described. The acute phase of infection

Figure 9.3 Scotty showing signs of dehydration and depression. (From Hartmann K & Levy J [2011] Feline Infectious Diseases. Courtesy of CRC Press.)

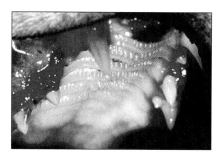

Figure 9.4 Scotty showing evidence of severe gingivitis. (From Hartmann K & Levy J [2011] Feline Infectious Diseases. Courtesy of CRC Press.)

is characterized by generalized lymphadenopathy, neutropenia, and fever. During the acute infection there is massive CD4+ T-cell depletion in all lymphoid tissues, especially the gut-associated lymphoid tissue (GALT) in the gastrointestinal tract. An asymptomatic period may then follow and last for months or years. Ultimately a persistent generalized lymphadenopathy occurs. This is accompanied by weight loss, leucopenia, opportunistic infection, recurrent fever, and glomerulonephropathies. The final phase of FIV infection (FAIDS), which is similar to AIDS in humans, is characterized by marked emaciation, tumors, neurological disease, and anemia.

COMPARATIVE MEDICINE CONSIDERATIONS

There are many similarities between human immunodeficiency virus and feline immunodeficiency virus. Both are lentiviruses that infect and deplete CD4+ T lymphocytes. Both have an acute phase, a relatively quiescent prolonged asymptomatic phase, and a final immunodeficiency phase that generally results in fatal opportunistic infections. Medications that depress HIV replication in human patients are now available. Many patients are treated with a cocktail of drugs that include protease inhibitors, reverse transcriptase inhibitors, fusion inhibitors, and entry inhibitors. Unfortunately, at the time of writing, such drugs are not available to treat FIV-infected cats.

Questions

1. Why can a persistent antibody response to FIV be used to diagnose infection with the virus?

2. Although the primary target cell for FIV is the CD4+ T lymphocyte, there is also a depression of innate immune responses. Explain why this occurs.

3. Glomerulonephritis is often seen in FIV-infected cats. Why might this be?

4. Describe the key similarities and differences between FeLV (see Case 8) and FIV infection in cats.

Further Reading

Hartmann K (2011) Clinical aspects of feline immunodeficiency and feline leukemia virus infection. *Vet Immunol Immunopathol* 143:190–201.

Lehman TL, O'Halloran KP, Hoover EA & Avery PR (2010) Utilizing the FIV model to understand dendritic cell dysfunction and the potential role of dendritic cell immunization in HIV infection. *Vet Immunol Immunopathol* 134:75–81.

Mohammadi H & Bienzle B (2012) Pharmacological inhibition of feline immunodeficiency virus (FIV). *Viruses* 4:708–724.

Policicchio BB, Pandrea I & Apetrei C (2016) Animal models for HIV cure research. *Front Immunol* 7:12.

Tompkins MB & Tompkins WA (2008) Lentivirus-induced immune dysregulation. *Vet Immunol Immunopathol* 123:45–55.

CASE 10
BOVINE VIRAL DIARRHEA VIRUS

Viruses that cause immunosuppression in the host usually target one or more cells of the immune system. Some of these viruses, such as human immunodeficiency virus (HIV) and feline immunodeficiency virus (FIV), produce disease as a result of the immunosuppression that they create. Some cause specific disease syndromes and then initiate or enhance other disease processes due to their immunosuppressive effect on the host. Fetal versus adult infection with a virus can also produce different effects on adaptive immunity. The varying clinical manifestations seen in cattle infected with bovine viral diarrhea virus (BVDV) (genus, *Pestivirus*; family, Flaviviridae) are linked to a number of factors, including the particular strain of infecting virus, the immune status of the host, and the age of the animal at the time of infection. Two types of BVDV have been identified based on virus behavior in cell culture, namely cytopathic and noncytopathic. Superimposed on this classification are two genotypes of BVDV, namely 1 and 2.

BVDV gains entry to macrophages and lymphocytes by binding to the CD46 molecule on the cell surface. This causes lymphopenia, apoptosis of lymphocytes in the thymus, and suppression of macrophage phagocytic function. The effects of BVDV on the host's immune system are different for noncytopathic (ncp) and cytopathic (cp) strains. This is particularly important with regard to apoptosis. Induction of apoptosis via the intrinsic pathway is documented for ncp BVDV, and the extrinsic pathway also possibly results in apoptosis of interfollicular T cells in the gut-associated lymphoid tissue (GALT). BVDV infects and kills lymphoid cells, particularly the ileal and jejunal Peyer's patches (Figure 10.1). Calves infected with BVDV will show lymphopenia and reduced responses to T-cell mitogens. B-cell apoptosis in follicular areas of the GALT has been attributed to reduced levels of Bcl-2, an anti-apoptotic factor. The development of oxidative stress between the host cell and cp BVDV is associated with cell death.

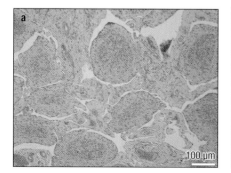
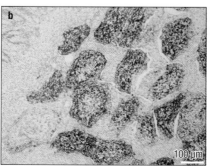

Figure 10.1 Histopathology of ileum from a BVDV-infected steer. (a) Section of ileum showing depletion of lymphocytes from Peyer's patches. (b) Section of ileum with immunohistochemistry staining to show the presence of BVDV in the tissue. (Courtesy of Dr. Mark Anderson, California Animal Health & Food Safety Laboratory, Davis, CA.)

The least complex clinical scenario for BVDV infection of naive adult cattle; such cattle will generally mount an effective immune response to BVDV and clear the infection. These infections can be classified as inapparent infections. Transient or acute infections are acquired after birth. Approximately 95% of postnatally acquired infections are of this type. Acute BVDV infection is often associated with leukopenia due to lymphopenia and neutropenia, as well as reduced innate and cellular defense mechanisms involving lymphocytes, neutrophils, and macrophages; together, such changes may predispose the animal to secondary infections that lead to conditions such as mastitis, metritis, and pneumonia. Other more severe clinical manifestations may also develop, and are associated with fever with diarrhea and severe anorexia, oral erosions, and coronitis. Death may occur in these cases.

Acutely infected cattle can spread the virus through several common mechanisms, including contact, flies, and aerosolization of virus. Cattle that are not pregnant and are not immune to BVDV develop a transient viremia which lasts for approximately 2 weeks after infection. Cows infected with BVDV while pregnant may abort if infection occurs early in the pregnancy. Calves that become infected *in utero* during early to mid-gestation are at high risk of becoming persistently infected with the disease.

THE CASE OF DWARF COW: A CALF WITH CHRONIC INFECTIONS

SIGNALMENT/CASE HISTORY

Dwarf Cow is a 9-month-old female Red Angus calf (**Figure 10.2**), born prematurely and bottle-fed by her owner. She is part of a small privately owned herd. One month ago she was treated for infectious bovine keratoconjunctivitis (pinkeye) with oxytetracycline and penicillin, and had apparently recovered. Recently the calf was noted to be lethargic, and her owner became concerned that she might have a respiratory infection.

PHYSICAL EXAMINATION

On physical examination, Dwarf Cow was normothermic with tachycardia (heart rate, 90 beats/minute; normal range, 60–80 beats/minute) and tachypnea (respiratory rate, 72 breaths/minute; normal range, 20–40 breaths/minute). No cardiac murmurs were auscultated. The calf weighed 51 kg and her body condition was poor; she showed increased respiratory effort and coughed frequently during the examination, indicating the likelihood of a respiratory infection with one or more bacterial pathogens. Thoracic auscultation revealed increased tubular airway sounds that were louder on the left hemithorax. Conjunctivitis was present, along with bilateral central corneal opacities with corneal neovascularization.

DIFFERENTIAL DIAGNOSIS

The most notable findings on physical examination were the extremely small frame of the calf and the increased respiratory effort with abnormal lung sounds. Common causes of poor body condition in calves include chronic bacterial or viral infections, internal abscessation, endoparasitism, trace mineral deficiencies (for example, selenium or copper deficiency), systemic organ disease (for example, renal or liver disease), and congenital cardiac defects. The abnormal lung sounds were probably due to bacterial causes secondary to a primary viral pneumonia; bacterial causes considered included *Mannheimia haemolytica*, *Pasteurella multocida*, *Mycoplasma* species, and *Histophilus somni*. Viral causes of pneumonia in young calves include

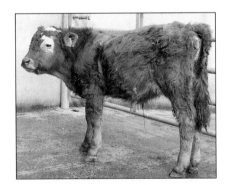

Figure 10.2 Dwarf Cow has been underweight since birth, and has been treated for numerous infections throughout her short life. (From Scott P, Penny C & Macrae A [2011] Cattle Medicine. Courtesy of CRC Press.)

infectious bovine rhinotracheitis (IBR) virus, bovine respiratory syncytial virus (BRSV), parainfluenza (PI3), and BVDV. The most likely chronic viral infection that could have accounted for the calf's small frame size was persistent infection with BVDV. The ocular changes were most probably associated with the previously diagnosed and treated *Moraxella bovis* infection (infectious bovine keratoconjunctivitis).

DIAGNOSTIC TESTS AND RESULTS

The initial diagnostic plan included a complete blood count (CBC), serum chemistry panel, whole blood selenium, trace mineral panel, thoracic radiographs, and abdominal and thoracic ultrasound. The CBC revealed an elevated total white blood cell count (13,780 white blood cells/μL; normal range, 5800–6800/μL) with monocytosis (1282/μL; normal range, 0–900/μL). The plasma protein concentration was low (5.7 g/dL; normal range, 6.5–8.5 g/dL) and the fibrinogen level was normal (300 mg/dL; normal range, 300–700 mg/dL). The chemistry panel revealed a low serum total protein concentration (5.0 g/dL; normal range, 7.0–8.7 g/dL) with a low globulin level (1.8 g/dL; normal range, 2.9–5.1 g/dL). Thoracic radiographs revealed an alveolar opacity within the caudoventral pulmonary parenchyma, and a moderate interstitial pattern in the perihilar region; an area of interstitial opacity was also present in the caudodorsal lung fields. Overall, the radiographic impression was consistent with pneumonia and caudodorsal pulmonary inflammatory infiltrates. Ultrasound examination of the right hemithorax revealed abscesses in the ventral pulmonary parenchyma at the level of the fifth and sixth intercostal spaces. In addition, approximately 11 well-encapsulated abscesses were identified throughout the abdomen. The kidneys and liver appeared normal. During the ultrasound examination the owner was consulted and consented to an ultrasound-guided aspiration of one of the pulmonary abscesses that appeared to be attached to the body wall. Thick tan pus was obtained; this material was submitted for aerobic and anaerobic culture and sensitivity and for *Mycoplasma* culture.

The most striking feature of the test results in this case was the presence of severe internal abscessation involving the lung and abdomen without increases in either fibrinogen or globulin. In cattle with internal abscesses, white blood cell counts are often normal because such abscesses are usually walled off. However, increased fibrinogen or globulin levels, or both, are typically observed. Given the generally small body frame and apparently reduced immune responsiveness in this calf, persistent infection (PI) with BVDV seemed likely, and an ear notch was collected and submitted for BVDV antigen-capture enzyme-linked immunosorbent assay (ELISA). Antibody assays in acutely infected cattle will usually be positive; PI animals will be negative for antibody but positive for virus/antigen. The technique of using formalin-fixed paraffin-embedded tissue from ear notch samples is a high-throughput method for identification of PI animals by immunohistochemistry. Ear notch samples are also used for antigen-capture ELISA and polymerase chain reaction (PCR) diagnosis of BVDV infection.

The results of the selenium and trace mineral panel were within normal ranges for cattle. The aerobic culture of the aspirated pus subsequently yielded mixed flora, including *Helococcus ovis*, based on partial DNA sequencing. The anaerobic culture was also mixed, and included *Bacteroides* species and *Prevotella* species that were β-lactamase negative. The *Mycoplasma* culture was negative. These are not typical organisms for causation of bovine pneumonia.

DIAGNOSIS

The result of the BVDV ear notch antigen-capture ELISA was positive, confirming a diagnosis of bovine viral diarrhea virus, most probably as a persistent infection. The BVDV ear notch test is most often used to determine whether cattle are persistently infected with BVDV. However, there have been reports of acutely infected cattle that are transiently positive on the ear notch ELISA. An acute infection results from postnatal infection, and thus the calf is not tolerant of the virus, so it is possible that the latter will be eliminated by the immune system. It was recommended that the owner should have this calf tested by BVDV reverse transcription polymerase chain reaction (RT-PCR) on whole blood in 3–6 weeks to confirm the suspected persistent BVDV infection. If persistently infected, the animal would remain positive for BVDV throughout its lifetime, as PI cattle do not revert to non-shedding status.

TREATMENT

There is no specific therapy for BVDV infection. In cases of acute BVDV, supportive fluid and electrolyte therapy coupled with antibiotic treatment (to prevent secondary bacterial infections) and nonsteroidal anti-inflammatory therapies are recommended. In this calf, antibiotic therapy with florfenicol was initiated while awaiting the results of the BVDV ear notch ELISA. Following confirmation of the positive BVDV ear notch ELISA, the owner was advised that this calf, given its likely PI status, would be a permanent source of BVDV exposure to other cattle, and would probably continue to suffer from recurrent or chronic bacterial infections. It was recommended that, if a subsequent BVDV test in 3–6 weeks confirmed BVDV, the calf should be culled or at least isolated from the rest of the herd. Since Dwarf Cow is part of a small private herd, if the owner wished to keep her it would be essential that the animal be kept segregated from all other cattle for the remainder of her life.

BOVINE VIRAL DIARRHEA VIRUS (PI FORM)

The PI form of BVDV observed in this case illustrates a complex interaction between the virus and its infected host. First, noncytopathic BVDV infects the developing fetus at between approximately 40 and 125 days of gestation. This early infection leads to a state of immune tolerance, wherein the host immune system identifies the infecting BVDV strain as self. Development of self/nonself recognition occurs in the thymus, and immune recognition of viral antigens as foreign can be masked if infection occurs during the critical window when positive and negative T-cell selection is occurring. During gestation, T lymphocytes develop in the thymus, where they acquire their α/β T-cell receptors. These receptors are screened for specificity to antigens presented by thymic epithelial cells in the thymus (**Figure 10.3**). Normally, T cells with receptors that recognize expressed antigens are purged (that is, undergo apoptosis) to prevent immune responses to self-antigens. If a viral antigen is present during gestation before development of immune competency (in this case, the 40- to 125-day window mentioned earlier), the fetal calf develops tolerance of the viral antigens, and the virus is able to persist in the animal throughout gestation and life. Following birth, calves from such a pregnancy serve as reservoirs for BVDV in cattle herds, as they are abundant shedders of the virus. Less than 5% of BVDV infections are of this type. A PI calf may be weak and die at birth, or it may appear normal and grow to maturity. Alternatively, an apparently healthy PI calf may succumb to a more severe form of disease before reaching adulthood if it is superinfected with cytopathic BVDV (mucosal disease can be a fatal sequela). Such infections typically lead to a fatal form of BVDV infection that may be acute, subacute, or chronic. Affected animals typically have fever, diarrhea, weight loss, gastrointestinal mucosal ulcers, and coronitis.

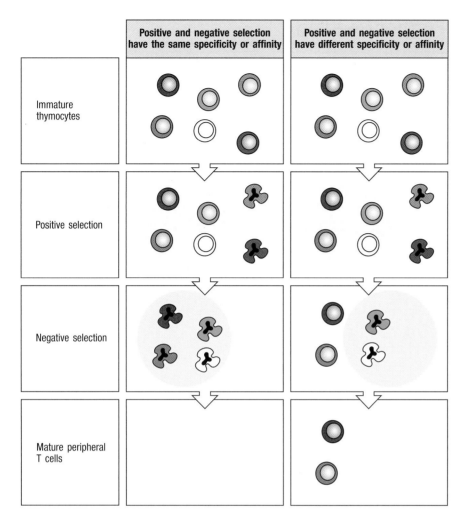

| Positive and negative selection have the same specificity or affinity | Positive and negative selection have different specificity or affinity |

Immature thymocytes

Positive selection

Negative selection

Mature peripheral T cells

Figure 10.3 Immature T cells are selected so that only those thymocytes with receptors that can engage peptide–MHC complexes on thymic epithelium survive to become MHC-restricted T cells. Negative selection removes T cells with receptors that are activated by self-peptides to produce self-tolerant T cells. The presence of viral peptides on cell membranes during this crucial stage of development causes the removal of cells with receptors that bind the viral antigens, and thus produces tolerance of the virus as though it were a self-antigen. (From Murphy K [2011] Janeway's Immunobiology, 8th ed. Garland Science.)

COMPARATIVE MEDICINE CONSIDERATIONS

In humans no similar virus has been described that causes development of tolerance during intrauterine infection, and then causes lymphocyte depletion after infection of the adult. There are viruses in humans and other species that replicate in lymphocytes and cause depletion of important immune cell populations (for example, HIV and FIV). The result of lymphocyte depletion is secondary infection of the patient with viral and bacterial pathogens.

Questions

1. What cells and tissues are targeted in BVDV-infected calves and how does the viral infection of these cells influence the immune response?

2. Explain how the timing of intrauterine infection with BVDV is critical in determining the effect of BVDV on the immune system and the potential development of persistent infection.

3. How does the ear notch assay diagnose BVDV infection?

Further Reading

Brodersen BW (2014) Bovine viral diarrhea virus infections: manifestations of infection and recent advances in understanding pathogenesis and control. *Vet Pathol* 51:453–464.

Lanyon SR, Hill FI, Reichel MP & Brownlie J (2014) Bovine viral diarrhoea: pathogenesis and diagnosis. *Vet J* 199:201–209.

Piccinini R, Luzzago C, Frigerio M et al. (2006) Comparison of blood non-specific immune parameters in Bovine virus diarrhoea virus (BVDV) persistently infected and in immune heifers. *J Vet Med B Infect Dis Vet Public Health* 53:62–67.

Van Metre DC, Tennant BC & Whitlock RH (2008) Infectious diseases of the gastrointestinal tract. In Rebhun's Diseases of Dairy Cattle, 2nd ed. (Divers TJ & Peek SF eds), pp. 258–273. Saunders.

CASE 11
MAREK'S DISEASE

The avian immune system functions in a similar fashion to the mammalian immune system. Following antigen stimulation and transformation, B lymphocytes generate specific antibodies and immunological memory. T lymphocytes are responsible for cell-mediated immunity, and also play a role in maturation of B cells (for example, helper T cells) and destruction of virally infected cells (for example, cytotoxic T cells). Neoplastic diseases of the avian immune system can be of infectious or non-infectious etiology. One infectious cause of immune system neoplasia is Marek's disease virus (MDV), a highly contagious epizootic herpesvirus, specifically gallid herpesvirus 2 (GaHVV-2). This pathogen is considered endemic in the global "poultry environment." There are multiple MDV-related syndromes, including lymphoproliferative syndromes, fowl paralysis, and skin leukosis.

MDV infects cells of the feather follicle, and can remain viable in feather dander for several months. The viable virus can be inhaled by susceptible chickens via desquamated feather follicle epithelium in poultry house dust, causing infection of B cells and ultimately transformation of T cells. This spread of the virus from cell to cell *in vivo* is facilitated by infected lymphocytes and epithelial cells.

Figure 11.1 Beaker showing signs of flaccid paralysis. (Courtesy of Maurice Pitesky.)

THE CASE OF BEAKER AND FRYER: CHICKENS IN A FAMILY FLOCK WITH PARALYSIS AND DROPPED WINGS

SIGNALMENT/CASE HISTORY

The Sanders family has a backyard chicken flock of 10 birds of varying breeds and mixed ages. The chickens have been acquired over the years from various sources, including feed stores, mail-order hatcheries, and a neighbor. The Sanders collect eggs daily and share them with neighbors and friends. Recently, they noticed that a 10-week-old pullet named Beaker had paralysis in one leg, weakened or dropped wings, and was in poor body condition for 2 days prior to death (Figure 11.1). Approximately 2 weeks later, a 5-month-old hen named Fryer developed paralysis and breathing difficulties before dying. In order to protect the remaining 8 birds in their flock, the owners brought Fryer to their small animal veterinarian. The referring DVM submitted the bird to the state diagnostic laboratory for necropsy. Many US states offer free or heavily discounted necropsy services for backyard poultry as part of their Avian Influenza and Exotic Newcastle Disease Surveillance Programs. According to the owner, the remainder of the flock appeared normal.

TOPICS BEARING ON THIS CASE:

Infectious oncogenic virus

Lymphoproliferative disease

Figure 11.2 (a) Pathology photo for Fryer, showing severe enlargement of the brachial plexi. The gross appearance of enlarged peripheral nerves, including the brachial plexi and sciatic nerves, is commonly considered pathognomonic for Marek's disease (MD). (b) Visceral lesions in the same hen, including tumors of the kidney, milliary nodules in the liver, and splenomegaly, can also be observed grossly. (Photographed by Gabriel Senties. Copyright © The Regents of the University of California, Davis Campus, 2011. All rights reserved. Used with permission.)

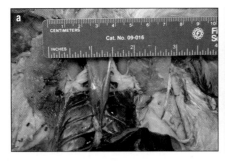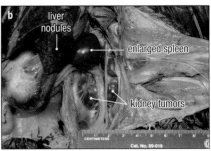

PHYSICAL EXAMINATION

The pathologist noted that Fryer had severe splenomegaly and marked thickening of the sciatic nerves and brachial plexi (**Figure 11.2A**). She also noted multifocal nodules in the liver and kidney, and enlargement of the spleen (**Figure 11.2B**). The gross appearance of the tumors and neural lesions, coupled with the signalment, were suggestive of Marek's disease (MD). Histopathologic examination of nodules revealed that they were composed of neoplastic lymphoid infiltrates (**Figure 11.3**).

DIFFERENTIAL DIAGNOSIS

Diagnosis of lymphoproliferative diseases in chickens and other poultry can be challenging due to the fact that there are multiple etiological agents capable of causing similar tumors. When gross neural lesions are found, a tentative diagnosis of MD is often made. However, thickening of peripheral nerves can have other causes, such as polyneuritis. Furthermore, a diagnosis of MD may be uncertain when tumors are present in the absence of peripheral nerve enlargement. In such situations it becomes necessary to examine both the nerves and the neoplastic tissue histologically. Indeed, the majority of MD diagnosis is based on signalment, gross pathology, and histology. However, even using these measures a diagnosis of MD cannot be made with 100% certainty. For example, the presence of nodules in visceral organs may suggest MD, lymphoid leukosis, or reticuloendotheliosis. A more

Figure 11.3 Histology photo (200× magnification) of the peripheral nerve demonstrating marked infiltration by neoplastic lymphocytes. (Photographed by Gabriel Senties. Copyright © The Regents of the University of California, Davis Campus 2011. All rights reserved. Used with permission.)

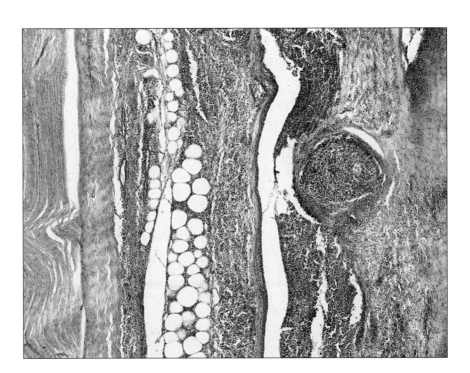

Table 11.1 Key features of Marek's disease (MD) for differentiation of this condition from lymphoid leukosis and reticuloendotheliosis

Age	Primarily affects birds less than 16 weeks old
Clinical signs	Wing and leg paralysis
Pathology	Affected chickens have peripheral nerve enlargement
	Peripheral nervous system involvement
	Lymphoid proliferation in skin and feather follicles
	T-cell lymphomas

definitive diagnosis of Marek's disease can be made by testing for markers of specific MD antigens in the neoplastic cells, using immunohistochemistry or immunofluorescence.

To help to clarify the diagnostic process, a step-wise methodology has been proposed for diagnosis of MD, consisting of (1) history, epidemiology, clinical observations, and gross necropsy, (2) characteristics of the tumor cell, and (3) virological characteristics[1] (Table 11.1).

DIAGNOSTIC TESTS AND RESULTS

Immunohistochemistry can be used to identify cell type and virus-specific antigens for MDV. However, since MDV is ubiquitous in poultry environments, isolation of the virus alone does not confirm a diagnosis of MD; the virus could be present in the bird without causing illness. This is why immunohistochemistry is a secondary diagnostic method after signalment, gross pathology, and histology have indicated the disease. Many laboratories are also now using the polymerase chain reaction (PCR) to confirm the diagnosis. Neither immunohistochemistry nor PCR were performed in this case.

DIAGNOSIS

Based on the signalment and gross and histologic lesions, Fryer was diagnosed with Marek's disease. Based on Fryer's diagnosis and Beaker's signalment, Beaker was believed to have also had MD.

TREATMENT

There is no treatment for Marek's disease. Therefore prevention is essential for the health both of individual birds and of the flock as a whole.

MAREK'S DISEASE (MD)

MD is often characterized by infiltration of pleomorphic neoplastic T cells, resulting in thickening of peripheral nerves and/or lymphomas and lesions in multiple tissues and organs, including liver, kidney, spleen, gonads, peripheral nerves, skin, and eye. The disease can also cause impairment of T-lymphocyte function, affecting both cell-mediated immunity and humoral immunity, and resulting in immunosuppression.

MDV can be controlled and largely prevented in poultry populations. Vaccination for MD constitutes an outstanding example of successful disease control in veterinary medicine. However, because the virus is ubiquitous in the environment, infection of chicks can occur almost immediately after hatching. The ideal time for vaccination is therefore either *in ovo* on the 18th day of incubation or by subcutaneous administration of vaccine at 1 day of age. This recommendation, coupled with the need to store the cell-associated vaccines

---

[1] Methodology developed by Witter et al. (2010).

in liquid nitrogen, makes vaccination for backyard breeders challenging. Therefore, for backyard operations that produce chicks, the lyophilized herpesvirus of turkey (HVT) vaccine, which is closely related to MDV, can be easily reconstituted and given subcutaneously at 1 day of age. Although this is generally the only viable vaccine option for producers who hatch their own eggs, lyophilized HVT vaccine is the least effective of the MDV vaccines and will not offer protection against more virulent strains of MDV. Due to the neutralizing effect of maternal antibody on the HVT vaccine, it can be ineffective in chicks (depending on the vaccination status of the hen).

Regardless of vaccination status, it is essential to place 1-day-old chicks in houses that have been thoroughly decontaminated to allow vaccinated birds time to develop immunity, which typically takes about 2 weeks. Ideally, backyard farms should be operated in an all-in all-out cycle in order to allow thorough cleaning and disinfection of the poultry environment to remove all traces of MDV before placement of a new flock. Although this may not be possible in some cases, veterinarians should work with their clients to set up a reasonable alternative with the goal of reducing the viral load of MDV in the housing environment.

COMPARATIVE MEDICINE CONSIDERATIONS

MD vaccines are the first effective vaccines against cancer in any species. In recent years a vaccine for human papillomavirus has been shown to be effective in prevention of cervical cancer when administered to young girls. The first commercial DNA vaccine was developed and is being used therapeutically for treatment of melanoma tumors in dogs and horses (off label). The melanoma vaccine does not contain a virus, but rather it utilizes the principle of cross-reactivity to activate an immune response. The enzyme tyrosinase is overexpressed in tumor melanoma cells, and by introducing a copy of the gene that encodes human tyrosinase in the DNA plasmid into the animal's skin, an immune response is activated against the melanoma cells.

Questions

1. Explain to the owner of the flock described in this case what, if anything, can be done to protect the remainder of the current flock.

2. The owners are considering buying some new chicks from a hatchery or feed store. What advice could you give them about the purchase and subsequent introduction of the new birds into their facilities?

3. Describe the route of infection of Marek's disease and how the virus typically causes disease.

4. If an owner has hens that were vaccinated using the HVT vaccine, what vaccine recommendation can you give for hatching eggs from those hens?

Further Reading

Fadly AM (2008) Neoplastic diseases. In Diseases of Poultry, 12th ed. (Saif YM ed.), pp. 449–512. Blackwell Publishing Australia.

Gennart I, Coupeau D, Pejaković S et al. (2015) Marek's disease: genetic regulation of gallid herpesvirus 2 infection and latency. *Vet J* 205:339–348.

OIE (World Organisation for Animal Health) (2013) Marek's disease. In: Manual of Diagnostic Tests and Vaccines for Terrestrial Animals 2013, pp. 566–575. OIE.

Randall C (1991) Diseases and Disorders of the Domestic Fowl and Turkey, 2nd ed. Mosby.

Shane SM (2005) Handbook on Poultry Diseases. American Soybean Association.

Witter RL, Gimeno IM, Pandiri AR & Fadly AM (2010) Tumor Diagnosis Manual: The Differential Diagnosis of Lymphoid and Myeloid Tumors in the Chicken. American Association of Avian Pathologists.

CASE 12
INFECTIOUS BURSAL DISEASE

The bursa of Fabricius is an epithelial and lymphoid organ that develops as an outpouching of the cloaca in birds (Figure 12.1). It was first described by Girolamo Fabrizi d'Acquapendente, Professor of Surgery at the University of Padua, in the early 1600s. The bursa is a sac-like structure with multiple internal folds containing follicles that each have a defined cortex. During ontogeny these follicles are populated with lymphoid stem cells that travel to this location from the fetal liver (Figure 12.2). Later in development these cells become lymphocytes (B cells) at different stages of maturation.

The discovery and study of the bursa contributed to delineation of the two main arms of the adaptive immune system. The bursa is the site of B-lymphocyte maturation in birds and the structure responsible for development of humoral immunity, while the thymus is the site of T-cell maturation and the major site for development of cell-mediated immunity. The bursa regresses with the onset of sexual maturity. Surgical removal of the bursa (bursectomy) from young chickens renders them unable to mount an antibody response to exogenously administered antigen. The bursa can also be affected by immunosuppressive viruses, such as the infectious bursal disease virus (IBDV).

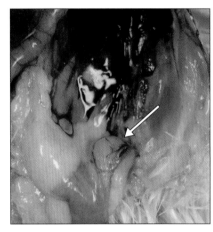

Figure 12.1 Anatomic location of the bursa of Fabricius (indicated by arrow) in a 21-day-old specific-pathogen-free (SPF) leghorn chicken. Note the location dorsal to the cloaca and ventral to both kidneys. (Courtesy of Rodrigo A. Gallardo.)

Figure 12.2 A healthy bursa of Fabricius follicle from a 28-day-old leghorn chicken. The normal histology of the follicle shows dense accumulation of lymphocytes and clear differentiation between the cortex (border of the follicle) and medulla. (Courtesy of Rodrigo A. Gallardo.)

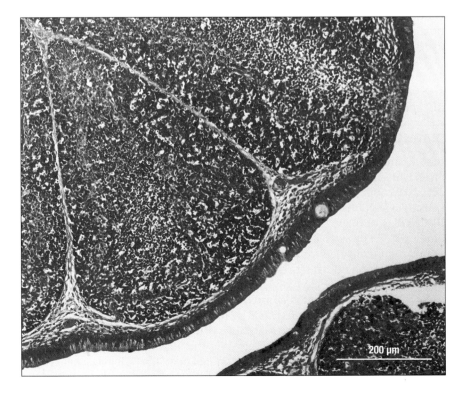

TOPICS BEARING ON THIS CASE:

B-lymphocyte development

Virus-induced immunosuppression

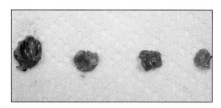

Figure 12.3 A normal bursa (shown on left) and three atrophied bursas. (Courtesy of Rodrigo A. Gallardo.)

THE CASE OF THE ROSS 308 FLOCK: A FLOCK OF BROILERS WITH A SUDDEN SPIKE IN DEATH RATES

SIGNALMENT/CASE HISTORY

A flock of 5-week-old broilers (commercial line Ross 308) have exhibited a mortality rate of 0.4% a day (4 chickens from a small flock of 1000 organic chickens) over the past few days. The farmer raises organic chickens four times a year. He processes and sells his birds at the local farmers' market. During processing he has found internal lesions, including air sac infection and hepatitis, in the carcasses of the birds that have recently died. The birds with lesions have been condemned and have not been sold as meat.

This is the first time that the farmer has had this problem. He has been buying his chicks from an established hatchery. At 1 day of age the chicks in the hatchery receive a vaccine against Marek's disease virus (MDV). At the farm, at 21 days of age they receive vaccines for infectious bronchitis virus (IBV) and Newcastle disease virus (NDV). The breeders from which the diseased birds were obtained were a 120-week-old flock, which was a different flock to the one from which the farmer had received his chickens previously. These breeders were given a killed IBDV vaccine at 18 weeks of age; they were not given live vaccine priming or killed booster.

PHYSICAL EXAMINATION

At necropsy, several birds (10/20) exhibited severe airsacculitis and coccidiosis, and one bird had ascites and perihepatitis. All of the chickens (20/20) had severely atrophied bursas (**Figure 12.3**). In most of the birds (15/20) the liver was enlarged with necrotic foci.

DIFFERENTIAL DIAGNOSIS

Chickens with airsacculitis could have NDV, IBV, *Mycoplasma gallisepticum* (MG), or inflammation due to poor ventilation (ammonia) and subsequent contamination by *Escherichia coli*. The presence of coccidiosis in conjunction with airsacculitis could be an indication of immunosuppression caused by IBD, *Bordetella avium*, chicken anemia virus (CAV), Marek's disease virus (MDV), stress caused by poor management, or exposure to mycotoxins. The small bursa is an indication of either IBD or MDV. Lesions in the liver would also be consistent with inclusion body hepatitis (IBH) (**Figure 12.4**), aflatoxins, and/or CAV.

DIAGNOSTIC TESTS

Blood samples were obtained from other birds in the flock in order to perform an enzyme-linked immunosorbent assay (ELISA) to look for antibodies against IBD, NDV, and IBV. Hemagglutination inhibition (HI) tests were performed for IBV and NDV (**Table 12.1**). The results showed low levels of antibodies against IBD, NDV, and IBDV. A set of tissue samples (bursa, thymus, spleen, liver, and bone marrow) was collected and fixed in formalin for histopathology; fresh tissues were collected for viral isolation tests. Bursa and spleen were tested for viral isolation of IBD by chorioallantoic membrane inoculation in 11-day-old embryonated specific-pathogen-free (SPF) eggs. A variant IBD strain was isolated from the inoculated embryos and confirmed by reverse transcription polymerase chain reaction (RT-PCR). Tracheas and respiratory exudate were obtained for viral isolation of respiratory viruses (IBV, NDV, etc.) by inoculating 11-day-old embryonated SPF eggs via the allantoic cavity. IBV was isolated after the second passage in embryonated eggs. RT-PCR in tracheal swabs

Table 12.1 Titers hemagglutination inhibition (HI) tests for NDV and IBV M41, and ELISA for IBDV

NDV	IBV M41	IBDV
0	4	40
0	16	80
5	8	80
0	16	80
0	8	40

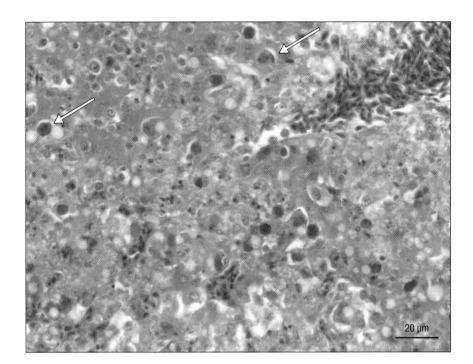

confirmed the presence of IBV and detected a lentogenic NDV strain. Liver histopathology revealed IBH; *E. coli* was isolated from liver and bone marrow. An adenovirus (primary cause of the IBH) was also detected in liver by PCR. Breeder serological information, specifically antibody titers for NDV, IBV M41, and IBDV (**Table 12.2**), was requested. The results showed non-uniform titers in breeders, specifically for IBDV. This is a common result of vaccinating breeders with a single inactivated vaccine without live vaccine priming.

DIAGNOSIS

Low and non-uniform IBDV antibody titers in breeders are suggestive of a lack of maternal antibodies in broilers, which leaves them unprotected against IBDV field challenge. This event, coupled with the isolation of a variant IBDV strain from the collected bursas, and the occurrence of secondary infections such as IBV, *E. coli*, and IBH, led to a diagnosis of immunosuppression caused by IBDV.

TREATMENT

There is no treatment other than supportive care and preventing the occurrence of the disease in subsequent flocks by implementing a better immunization strategy in breeders and/or broilers. Rigorous vaccination schedules in breeders to increase the delivery of maternal antibodies (passive immunity) to the progeny can help to prevent IBD. The choice of vaccines should reflect the strains that are prevalent in the region. Another strategy would be to vaccinate broilers and protect them by active immunity. Live attenuated, killed, and recombinant vaccines are commercially available for broiler and breeder vaccination protocols. Revaccination in older flocks might be an effective strategy, and should have been undertaken in this case, as the broilers came from a 120-week-old breeder flock that received a single killed vaccine at only 18 weeks of age. Overall, future implementation of good management practices and biosecurity was recommended to this owner. This included single-age flocks, vaccination protocols for coccidia, IBDV, and respiratory diseases monitored by serological tests, cleaning and disinfection, and an adequate down-time period between flocks. Since the birds in this case were raised organically, no treatment could be used to counteract

Table 12.2 Titers from 40-week-old breeder flock (HI for NDV and IBV-41, and ELISA for IBDV-1)

NDV		IBV-41		IBDV-1	
5	10	256	64	5	10
10	5	32	128	80	20
5	5	64	32	640	40
10	0	32	128	1280	1280
0	5	128	64	640	80

the coccidiosis and bacterial infections that were causing airsacculitis. The potential for significant economic loss for organic producers is high when immunosuppression causes infection.

INFECTIOUS BURSAL DISEASE

Infectious bursal disease is a highly prevalent disease in chickens. The control strategy has two steps: (1) hyperimmunization of breeders to provide a good level of maternal antibodies to the progeny, and (2) immunization of the progeny. In this case, breeders were not primed with a live IBD vaccine prior to receiving an inactivated vaccine, leaving the progeny (broilers) unprotected against IBDV. The absence of memory cells created by live vaccine priming resulted in breeders that responded poorly to a single inactivated vaccine. In the USA, variant strains of IBD are the most prevalent IBD strains that cause problems in young flocks. The IBDV affects the bursal follicles, inducing lymphoid depletion and subsequent immunosuppression. If the damage is severe, the affected bursal follicles will never be replenished with lymphocytes, leading to permanent immunosuppression that can be detected by secondary infections and poor response to vaccines applied as part of prevention of endemic diseases. As shown in this case, the immunosuppression created by an early IBDV infection opens the door for an opportunistic adenovirus that is able to cause IBH. The other viruses, such as the isolated IBV and NDV, were opportunistic and acquired from either vaccination or field exposure. Coccidiosis events were also linked to the immunosuppression.

COMPARATIVE MEDICINE CONSIDERATIONS

For many years the chicken was considered to be the only species in which natural infection with IBDV occurred. All chicken breeds are affected, and more severe clinical signs and mortality are observed in white leghorns than in broilers. IBDV and specific antibodies to it have now been detected in several bird species; in most of these birds, infection is not associated with pathology. There is really no equivalent of this disease in mammals, as they do not have a bursa of Fabricius, which is the target organ for this birnavirus.

Questions

1. What is the role of the bursa in immune system development in avian species? What would the result be if neonatal bursectomy was performed?

2. Why does infectious bursal disease virus increase the incidence of secondary infections in infected chickens?

3. How would you expect an IBDV-infected chicken to respond to a vaccination for another avian pathogen?

Further Reading

Adelmann HB (1942) The Embryological Treatises of Hieronymus Fabricius of Acquapendente. Cornell University Press.

Cooper MD, Peterson RDA & Good RA (1965) Delineation of the thymic and bursal lymphoid systems in the chicken. *Nature* 205:143–146.

Davidson F (2013) The importance of the avian immune system and its unique features. In Avian Immunology, 2nd ed. (Schat KA, Laspers B & Kaiser P eds.), pp. 1–10. Academic Press.

Glick B (1987) How it all began: the continuing story of the bursa of Fabricius. In: Avian Immunology: Basis and Practice (Toivanen A and Toivanen P eds.), pp. 1–8. CRC Press.

Glick B, Chang TS & Jaap RG (1956) The bursa of Fabricius and antibody production. *Poultry Science* 35:224–225.

Senne DA (2008) Virus propagation in embryonating eggs. In: A Laboratory Manual for the Isolation, Identification, and Characterization of Avian Pathogens, 5th ed. (Dufour-Zavala L, Swayne DE, Glisson JR et al. eds.), pp. 204–208. The American Association of Avian Pathologists.

CASE 13
BOVINE RESPIRATORY DISEASE COMPLEX

Stress can predispose animals to the development of viral and bacterial disease because the overproduction of cortisol depresses the immune system. Bovine respiratory disease (BRD) complex, a common and important cause of mortality and morbidity in both dairy and beef cattle, is an excellent example of the interaction of environmental stress with bacterial and viral pathogens in disease pathogenesis. Of the factors that contribute to the development of bovine respiratory disease complex, also called shipping fever, it is evident that at least some are sources of stress, particularly recent abrupt weaning, mixing with cattle from multiple origins, and shipment for a prolonged period of time. Anyone who has witnessed a young animal abruptly weaned will have observed the signs of separation anxiety that are often exhibited, including vocalization and restless walking or trotting about in search of the mother. Mixing with animals of multiple origins requires reorganization of the social hierarchy, leading to opportunities for conflict. Shipping on a livestock trailer is a stressful event to which young calves are not accustomed.

Although brief episodes of stress with cortisol release may actually improve immune responsiveness, sustained or severe stress suppresses immune function, and can lead to increased rates of infectious disease if animals are exposed to infectious agents during the same period. Shipment of recently weaned calves has been demonstrated to increase plasma cortisol levels and decrease the blastogenesis response of blood lymphocytes exposed to the mitogen phytohemagglutinin. Abrupt weaning has been associated with an increase in death rates when calves are exposed to bovine herpesvirus 1 (BHV-1) and *Mannheimia haemolytica.* This fatal pneumonia is also associated with increased production of the cytokine interleukin 10 (IL-10), which suppresses some T-lymphocyte functions. Because stress can have such a significant impact on rates of disease and death, many cattle contract buyers require the use of a practice called "preconditioning" to decrease the impact of stress on the health of calves.

THE CASE OF THE TRAVELING FEEDLOT CALVES: A GROUP OF CALVES WHO DEVELOPED FATAL RESPIRATORY DISEASE

SIGNALMENT/CASE HISTORY

A group of 24 Simmental crossbred calves ranging from 5 to 7 months of age were weaned from their mothers at the Georgia farm where they were born, and sent to a local auction market an hour's drive away. At the auction market they were purchased by an order buyer for a large Kansas feedlot. The order

TOPICS BEARING ON THIS CASE:

Effects of stress on immunity

Synergy of viral and bacterial agents in disease causation

Secondary immunodeficiency

Figure 13.1 The group of recently shipped calves showing signs of stress. (From Scott P, Penny C & Macrae A [2011] Cattle Medicine. Courtesy of CRC Press.)

buyer also purchased other calves at the same auction market until he had accumulated a group of 100 calves in total, all similar in age and breed to the original group of 24 Simmental crossbred calves, but having been born on various farms in Georgia and Tennessee.

The entire group was housed in several pens at the sale barn, and was provided with hay and water overnight. The next morning the animals were shipped on a livestock truck to the Kansas feedlot (**Figure 13.1**). Upon arrival in Kansas, the calves were treated with a dose of long-acting oxytetracycline and an injectable anthelminthic (ivermectin). They were also given a multivalent clostridial bacterin and a modified live vaccine containing bovine herpesvirus 1 (BHV-1), parainfluenza type 3 virus (PI3V), bovine respiratory syncytial virus (BRSV), and bovine viral diarrhea virus type 1 and type 2 (BVDV-1 and BVDV-2). The male calves were castrated, and several calves required dehorning. The 100 calves were put into a single pen and checked twice daily for any signs of illness. They were provided with free-choice water, grass hay, and 2 pounds per head per day of a 14% protein concentrate mix.

On day 7 after arrival, two of the calves were found dead in the pen, and five more calves were noted to be hanging back from the group and disinterested in feed.

PHYSICAL EXAMINATION

Necropsy of the two dead calves showed that they both had fibrinous pleuropneumonia (**Figure 13.2**), characterized by deposition of fibrin on the pleural surfaces, accumulation of viscous yellow fluid in the pleural space, and consolidation of the cranioventral lung lobes bilaterally. The five calves that appeared depressed were clinically examined. Rectal temperatures for the calves were elevated, in the range 104.4–105.7° F (normal range, 100.5–102.5°F); respiratory rates were also elevated (hyperpnea), in the range 66–84 breaths/minute (normal range 26-50 breaths/minute). One calf was heard to cough when it entered the chute for examination. Thoracic auscultation revealed large airway sounds over the cranioventral lung fields of three of the calves; it was not clear whether airway sounds were abnormal in the remaining two calves. Rumen fill was decreased in all of the calves, and feces passed by two calves in the chute seemed somewhat dry.

DIFFERENTIAL DIAGNOSIS

Respiratory disease in cattle can be caused by several viruses, including BHV-1, BRSV, PI3V, BVDV, and bovine corona virus. The bovine respiratory disease complex includes one or more of these viruses, as well as one or more of the following bacteria: *Pasteurella multocida*, *Mannheimia haemolytica*, *Mycoplasma bovis*, and *Histophilus somni*. Another condition that is a possible cause of respiratory disease in cattle is atypical interstitial pneumonia.

DIAGNOSTIC TESTS AND RESULTS

Diagnostic microbiological tests were not submitted in this case to confirm the exact microbiological cause of the disease for each calf, but the likely cause of infection was one or more viruses (BHV-1, BRSV, BVDV, PI3V, and bovine respiratory coronavirus are all possible candidates) and secondary infection with *Mannheimia haemolytica* (the bacterium most commonly associated with acute fibrinous pleuropneumonia in cattle).

DIAGNOSIS

Based on their clinical signs (along with necropsy findings for pen-mates), the five calves were presumptively diagnosed with bovine respiratory disease complex resulting in acute bronchopneumonia or pleuropneumonia.

Figure 13.2 Fibrinous pleuropneumonia identified on necropsy of one of two calves that died suddenly on day 7 after being introduced to the feedlot. (From Scott P, Penny C & Macrae A [2011] Cattle Medicine. Courtesy of CRC Press.)

TREATMENT

All five calves were treated with florfenicol by subcutaneous injection and returned to their pen. One of the five calves died the next day, but the remaining four animals improved over the next week and required no further treatment. Other calves in the pen showed similar clinical signs of depression and loss of appetite over the next 2 weeks. Each day, two to eight additional calves required examination; all were found to have fever and hyperpnea, with some calves coughing occasionally. By the end of the initial 2-week period at the feedlot, 68 of the 100 calves had been treated for acute pneumonia, and eight calves had died. At necropsy all of those animals were confirmed to have extensive fibrinous pleuropneumonia.

The supervising veterinarian and the feedlot manager discussed the outbreak, and agreed that it was not surprising that some members of this group of calves developed acute pneumonia, due to their history of passing through a sale barn and being shipped on a truck soon after weaning. However, they also noted that the outbreak seemed to be more virulent than expected, as the affected calves died rapidly, with few or no signs of disease prior to their death.

BOVINE RESPIRATORY DISEASE COMPLEX

Bovine respiratory disease complex is not due to just one cause, but rather to a combination of multiple factors, which converge to cause severe and possibly fatal pneumonia. At least two of the following factors are typically implicated when a group of cattle develops BRD: lack of protective immunological memory relating to respiratory pathogens; recent weaning; recent castration; dehorning; mixing with cattle from multiple origins; shipment over relatively long distances or for prolonged periods (1 hour or more); extremes of weather; abrupt introduction to a new diet; and exposure to one or more respiratory pathogens (usually from other cattle).

Although the calves in this case received a vaccine upon arrival at the feedlot, which theoretically could have helped them to develop protective immunity to respiratory viruses, they had by that stage been mingling together for several days, so it is likely that infected animals had already transmitted one or more respiratory viruses to other calves in the group.

In a preconditioning program, calves are weaned from their mothers 30–45 days in advance of shipping. During this period, the calves are given time to adapt to a new diet that no longer includes grass, or milk from their mothers, and they drink out of water troughs. They are also treated for parasites and administered a priming dose of vaccines against respiratory and other pathogens, along with appropriately timed boosters, so that they have time to develop protective immune memory. Castration and dehorning, if necessary, are completed early enough to ensure that the calves are completely healed before they are shipped anywhere or allowed to mingle with other cattle. If calves that have been preconditioned are not mixed with any other cattle before being shipped to a feedlot, or if they are only mixed with cattle that have been similarly preconditioned, morbidity and mortality rates associated with BRD are usually much lower.

COMPARATIVE MEDICINE CONSIDERATIONS

It is well known from data for humans and laboratory rodents that corticosteroids suppress immune function. Stress is a common inducer of adrenal corticosteroid hormone release. The role of the CD8$^+$ T cell in antiviral immunity is well established, and studies have demonstrated that both primary CD8$^+$ T-cell responses and generation of memory CD8$^+$ T cells are reduced by exposure to endogenous and/or exogenous corticosteroids. In addition, major histocompatibility complex (MHC) class I antigen processing by dendritic cells has been reported to be compromised by stress in laboratory rodents. This is an important mechanism for activation of T lymphocytes for eventual destruction of virus-infected cells.

Questions

1. Explain how a viral infection can predispose calves to secondary infection of the lung with bacteria.

2. How did the process of weaning, castration, and shipping facilitate the development of severe pneumonia in this case?

3. What serological tests could be performed on the surviving calves to determine which virus or viruses were involved?

Further Reading

Bailey M, Engler H, Hunzeker J & Sheridan JF (2003) The hypothalamic-pituitary-adrenal axis and viral infection. *Viral Immunol* 16:141–157.

Filion LG, Wilson PJ, Bielefeldt-Ohmann H et al. (1984) The possible role of stress in the induction of pneumonic pasteurellosis. *Can J Comp Med* 48:268–274.

Griebel PJ & Hodgson PD (2009) Abrupt weaning significantly increases mortality following a secondary bacterial respiratory infection. *Proc Am Assoc Bov Pract* 42:23–36.

Powell ND, Allen RG, Hufnagle AR et al. (2011) Stressor-induced alterations of adaptive immunity to vaccination and viral pathogens. *Immunol Allergy Clin North Am* 31:69–79.

Purdy CW, Richards AB & Foster GS (1991) Market stress-associated changes in serum complement activity in feeder calves. *Am J Vet Res* 52:1842–1847.

CASE 14
ALLERGIC ASTHMA

Allergic respiratory disease is caused by inhalation of environmental allergens. Professional antigen-presenting cells called dendritic cells reside below the respiratory airway epithelial cells and send projections (dendrites) to the airway lumen that enable them to capture inhaled antigens. Dendritic cells then process these antigens internally and present a portion of the antigen on their cell surface in conjunction with major histocompatibility complex (MHC) II. Local naive T-helper 0 (Th0) cells recognize the allergen in conjunction with MHC II, and are stimulated to become Th2 cells, which produce a variety of cytokines (proteins that serve as messages to other cells). Thus the Th2 cell is considered to be the major cell that orchestrates the subsequent inflammatory airway response. The cytokines produced by Th2 cells can induce the maturation, differentiation, and survival of eosinophils, and stimulate B cells to produce antibodies of the IgE class. IgE antibodies that are secreted become bound to mast cells just below the surface of the airways. Upon re-exposure to the allergen, two IgE antibodies are cross-linked, sending an intracellular message and triggering mast-cell degranulation (Figure 14.1).

Asthma in cats is believed to result predominantly from a type I hypersensitivity (that is, IgE-mediated) response to aeroallergens under appropriate genetic and environmental influences. Inhalation of pollen allergen causes the production of IgE antibodies, which bind to Fc epsilon receptors on mast cells in the respiratory tract. Upon a second exposure to the allergen, acute release of mast-cell mediators occurs. A variety of preformed mediators within mast-cell

TOPICS BEARING ON THIS CASE:

Type 1 (IgE-mediated) hypersensitivity

Differential activation of T-helper type 1 and T-helper type 2 cells

Mast-cell mediators and cytokines

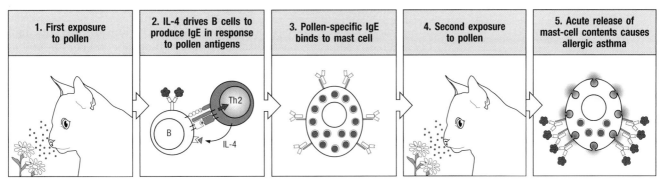

Figure 14.1 The cat inhales pollens (red particles) (frame 1), which bind to B-cell receptors and interact with antigen-presenting cells that then drive T cells to make the cytokine interleukin 4 (IL-4) (frame 2). This IL-4 strongly stimulates B lymphocytes to differentiate into plasma cells and make IgE antibodies. The pollen-specific IgE binds tightly to FcIgE receptors on the mast-cell membranes (frame 3). Upon re-exposure to the pollen (frame 4), antigen inhaled by the cat cross-links the IgE on the cell surface (frame 5). This triggers the release of potent mast-cell mediators by degranulation of the mast cell, and stimulates production of additional cytokines and leukotriene mediators, which cause the clinical signs of allergic asthma. (From Geha R & Notarangelo L [2016] Case Studies in Immunology, 7th ed. Garland Science.)

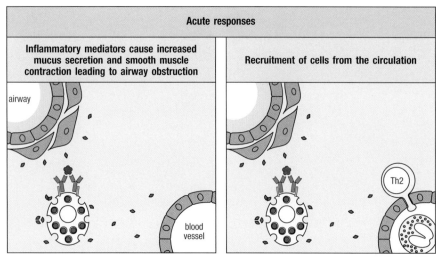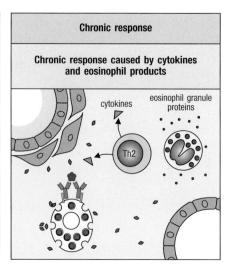

Acute responses		Chronic response
Inflammatory mediators cause increased mucus secretion and smooth muscle contraction leading to airway obstruction	Recruitment of cells from the circulation	Chronic response caused by cytokines and eosinophil products

Figure 14.2 Cross-linking of IgE molecules on mast cells by allergen (denoted by red star) causes the release of vasoactive mediators (denoted by blue dots). The preformed mediators, which include histamine, cause increased capillary permeability and smooth muscle contraction. These physiological effects result in increased respiratory secretions/mucus production and bronchoconstriction, which are responsible for the clinical signs in asthma. Chemotaxis of eosinophils to the airways occurs and contributes to airway inflammation. (From Murphy K [2011] Janeway's Immunobiology, 8th ed. Garland Science.)

granules are immediately released, causing bronchoconstriction, increases in vascular permeability. and an inflammatory cell influx. As other eicosanoids and cytokines are elaborated by the mast cells, additional cells infiltrate the airways, and a second wave of bronchoconstriction (the "late-phase response") can occur from hours to a couple of days later (Figure 14.2).

THE CASE OF SMOKEY: A CAT WITH CHRONIC ACUTE RESPIRATORY DISTRESS

SIGNALMENT/CASE HISTORY

Smokey, a 2-year-old spayed female domestic shorthaired cat (FS DSH) was presented to her veterinarian with an acute onset of respiratory distress. She had had three or four episodes of transient rapid breathing over the past 6 months, but these episodes were self-limiting. Her current episode started 2 hours ago and was described as initially rapid breathing. Over time, her breathing effort had become labored and had progressed to open-mouthed breathing.

The owner was questioned about other respiratory clinical signs, but had not noted any sneezing, nasal discharge, or coughing. She did comment that Smokey frequently hacked up hairballs; when more specifically questioned, the owner admitted that Smokey rarely produced a hairball (maybe once a month), but spent a lot of time (several times daily) trying to "get the hairball up." Smokey lives predominantly indoors and is the only pet in the home. The owner does not smoke, but she does burn both incense and candles. In addition, some aerosols are used (cleaning products, air fresheners, cooking sprays, and hairspray). No seasonality of the respiratory distress was noted. Smokey has no significant travel history, she is fully vaccinated, and since 6 months of age she has been receiving monthly heartworm preventive medication (selamectin).

PHYSICAL EXAMINATION

On physical examination, Smokey was found to have a respiratory rate of 68 breaths/minute (normal range, 20–30 breaths/minute), with a pronounced expiratory effort ("push"). Intermittently she extended her head and neck and would open her mouth to breathe. With gentle tracheal pressure, a cough was easily elicited; the owner confirmed that this cough was what she had thought was an effort to hack up a hairball. On auscultation, expiratory wheezes could be heard. No other abnormalities were noted.

DIFFERENTIAL DIAGNOSIS

Differential diagnoses for a cat with respiratory distress include upper airway obstruction, lower airway obstruction, pulmonary parenchymal disease (pneumonia, cardiogenic or non-cardiogenic pulmonary edema, neoplasia, interstitial lung disease, hemorrhage, etc.), pleural cavity disorder (air, fluid, or mass), chest wall or diaphragmatic defect or dysfunction, and pulmonary thromboembolism and "look-alike" disorders (for example, hyperthermia, acidosis, pain, drug effects). Given that Smokey showed respiratory distress during the expiratory phase of respiration, and that expiratory wheezes were auscultated, the differential diagnoses can be narrowed to causes of lower airway obstruction. This can suggest a narrowed diameter of the intrathoracic trachea (uncommon in cats) or inflammatory lower airway disease (common in cats).

DIAGNOSTIC TESTS AND RESULTS

As cats with respiratory distress are fragile patients, steps were immediately taken to stabilize Smokey. These included administration of a bronchodilator (terbutaline 0.01 mg/kg intramuscularly) and a glucocorticoid (dexamethasone 0.25 mg/kg subcutaneously once) to alleviate bronchospasm and reduce inflammation, respectively. Smokey was then placed in an oxygen cage to minimize stress, and respiratory rate and effort were repeatedly assessed (without handling her) every 5–30 minutes until she stopped open-mouthed breathing and her respiratory rate and effort improved. Thoracic radiographs were then taken, and showed a marked bronchial pattern with collapse of the right middle lung lobe and the caudal portion of the left cranial lung lobe (Figure 14.3).

The differential diagnoses for a bronchial pattern include feline asthma, lungworm infection (*Aelurostrongylus abstrusus*), heartworm-associated respiratory disease (HARD), and chronic bronchitis. Jugular venipuncture

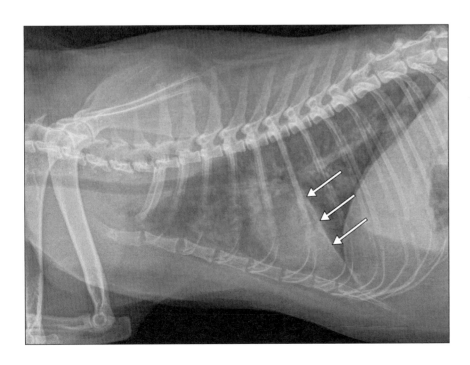

Figure 14.3 A lateral radiograph of Smokey's thorax showing atelectasis of the right middle lung lobe (indicated by arrows) and the caudal portion of the left cranial lung lobe. (Courtesy of Carol Reinero.)

was performed to obtain blood for a complete blood count, which revealed a peripheral eosinophilia (3400/uL; reference range, < 1500/uL). A fecal Baermann test was performed, which did not show evidence of parasites. Testing for HARD was not deemed necessary, as the owner was reliably administering heartworm preventive medication on a monthly basis. Terbutaline was administered every 4–6 hours (0.01 mg/kg subcutaneously) as needed when the cat's respiratory rate exceeded 40 breaths/minute or her respiratory effort was labored. The following day, without supplemental oxygen Smokey had a respiratory rate of 28 breaths/minute, and no expiratory "push" was noted. She was subsequently anesthetized for collection of bronchoalveolar lavage fluid (BALF) for cytology, culture, and polymerase chain reaction (PCR) for mycoplasma. Prior to anesthetic induction, she was given a dose of terbutaline (0.01 mg/kg subcutaneously). BALF cytology revealed 3500 cells/μL, with 67% eosinophils and 12% non-degenerate neutrophils (reference ranges, < 250 cells/μL, < 17% eosinophils, and < 7% non-degenerate neutrophils, respectively). The remainder of the cells were alveolar macrophages. There was no evidence of parasites in the lavage fluid. PCR for mycoplasma was negative, as were the aerobic and anaerobic cultures.

DIAGNOSIS

Smokey was diagnosed with chronic allergic asthma based on a combination of compatible clinical signs (cough and episodic expiratory respiratory distress), peripheral eosinophilia, thoracic radiographic features, negative tests for lungworm (Baermann test and BALF cytological examination), exclusion of HARD (due to consistent monthly treatment with selamectin), and evidence of BALF eosinophilia. The characteristic feature of chronic bronchitis is non-degenerate neutrophils in BALF, and although Smokey had increased airway neutrophil numbers, the eosinophil was the predominant cell type. In addition, chronic bronchitis is not associated with episodic respiratory distress, which is in part alleviated by administration of bronchodilators.

TREATMENT

There are currently three major ways in which chronic asthma is treated in cats, namely environmental modulation, bronchodilators, and glucocorticoids. Environmental modulation relies on decreasing any nonspecific irritants to the respiratory tract, including smoke and aerosols or dusts. Smokey's owner was encouraged to avoid using candles and incense or other aerosols, or if she did use them, to make sure that Smokey was in a different room. It was also recommended that she should switch Smokey's cat litter to a dust-free type. High-efficiency particulate air (HEPA) filters were advocated for the rooms where Smokey spent most of her time. The owner was also instructed to wash Smokey's bedding weekly in hot water, to increase the frequency with which she vacuumed (and to use a vacuum cleaner with a HEPA filter), and to stop using her feather duster (that Smokey liked playing with).

Avoidance of specific allergens implicated in allergic asthma is also helpful, if they can be identified by skin or serum testing, which the owner declined. Some of the previously described measures for nonspecific irritants would also decrease common indoor aeroallergens. Alternatively, the cat could be removed from its environment—for example, a cat that is allergic to indoor allergens could be made an outdoor cat, and vice versa.

Bronchodilators play a key role in reducing airflow limitation from bronchospasm, and can be administered orally, by injection, or by the aerosol route. Common classes of bronchodilators include methylxanthines and beta-2 agonists. The methylxanthines have a relatively narrow therapeutic index, and toxicity can be a concern without serum therapeutic monitoring. Regular use of inhaled racemic short-acting beta-2 agonists (for example, albuterol) has been linked to paradoxical bronchoconstriction and exacerbation of eosinophilic airway inflammation, and overuse should be avoided. In any case, bronchodilators should not be administered as monotherapy for the treatment of asthma, as they are ineffective in reducing airway inflammation, a key component of this disorder, which contributes to further airway hyperresponsiveness and airway remodeling (permanent architectural changes in the lung leading to a decline in lung function that is not responsive to medical management). Bronchodilators are of most use in cats that have episodes of respiratory distress, and may not be necessary for cats that present with cough alone. They are absolutely critical for the management of bronchoconstriction that occurs during an acute asthma attack. They may be given to the owner to take home for use as "rescue" therapy if an asthmatic attack occurs outside the hospital environment.

Glucocorticoids represent the mainstay of therapy for feline asthma, as they reduce airway eosinophilia and aid the control of clinical signs. Like bronchodilators, they can be administered orally, by injection, or by aerosol. Topical aerosol delivery of glucocorticoids is associated with fewer systemic effects on the immune and endocrine systems, and has become popular in feline medicine. They can be administered by use of a spacer and tight-fitting face mask. The cat should be watched for 8–10 deep respirations with the face mask in place. Administration of inhaled medications by waiting for a certain amount of time to elapse is not effective, as the cat may either hold its breath or take shallow breaths when the face mask is in place. The common inhaled glucocorticoids (for example, fluticasone and flunisolide) need to be administered twice daily. High doses may not be associated with an improved clinical benefit, as a plateau effect can be seen with lower doses. Inhalant glucocorticoids are not advocated for use in an asthmatic crisis, as they probably take several days to have an effect on asthmatic inflammation. In an asthmatic crisis, injectable glucocorticoids are generally administered. For chronic management, oral glucocorticoids (for example, prednisolone) are started and can be switched to inhalant steroids as indicated.

Smokey was sent home on oral prednisolone (10 mg/day), metered-dose inhaler albuterol as needed for "rescue" therapy (Figure 14.4), and the previously described environmental modifications. She returned for a recheck examination 3 weeks later. Her owner commented that Smokey had had no episodes of coughing or respiratory distress since she left the hospital. Thoracic radiographs showed marked improvement of her previously documented bronchial pattern. Her right middle lung lobe remained collapsed, but there was resolution of the atelectasis associated with the caudal portion of the left cranial lung lobe. The owner opted not to switch to inhaled steroids at that time. The dose of Smokey's prednisolone was subsequently tapered over the next few weeks, and she was maintained with minimal clinical signs on a dose of 2.5 mg/day along with occasional use of inhaled albuterol (1–2 times/month) as needed for clinical signs of bronchoconstriction.

ALLERGIC ASTHMA

Allergic asthma in the cat is similar to the disease described in humans. Many of the environmental allergens are identical. Details of the condition as it occurs in both species are described below.

COMPARATIVE MEDICINE CONSIDERATIONS

Allergic asthma is an important respiratory disease in humans. It often appears in young children and persists through adulthood, although adult-onset asthma also occurs. Asthma patients present with tightness in the chest as a result of bronchoconstriction, increased mucus in airway secretions, and a cough. Allergic asthma is caused by the production of IgE in response to allergens present in the environment. Synthesis of leukotrienes contributes to the more protracted clinical signs that are not mitigated by use of antihistamines.

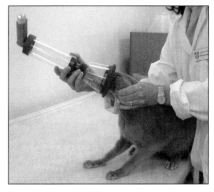

Figure 14.4 Smokey receiving albuterol via a metered-dose inhaler. This medication helps to dilate the airways and ease breathing. (From Schaer M [2009] Clinical Medicine of the Dog and Cat, 2nd ed. Courtesy of CRC Press.)

In chronic asthma in human patients there is extensive airway remodeling, with hyperplasia of mucus-producing epithelial cells.

Feline asthma presents in a very similar manner to human asthma. The mechanism also involves IgE formation and mediator release. In cats, both histamine and serotonin are important mediators. Pulmonary function studies on asthmatic cats show increased pulmonary resistance as a result of bronchoconstriction. The allergens that cause asthma in cats are often the same as those that stimulate IgE production in their human counterparts. House-dust mites, pollens, and avian proteins (from feather pillows or dusters) are common allergens.

In both cats and humans there is a genetic predisposition to development of asthma. This may be multifactorial, involving both IgE responsiveness and airways hypersensitivity. Methacholine challenge can be performed on cats and on human patients to evaluate the degree of hyperresponsiveness of airway smooth muscle. Many asthmatics are unable to tolerate perfumes, cigarette smoke, and other potential airway irritants.

Treatment of asthma in both species is similar, namely avoidance of the allergen when possible, anti-inflammatory therapy with injectable, oral, or inhaled corticosteroids, and "rescue" therapy with an inhaled bronchodilator.

Questions

1. Explain why Smokey developed difficulty breathing and a cough even though she was predominantly an indoor cat. What are some of the allergens that could have induced asthma in this patient?

2. The predominant cell type in the bronchoalveolar lavage was the eosinophil. Describe the immunological events that caused the peripheral eosinophilia as well as the large number of these cells in the respiratory secretions.

3. Smokey's owner did not wish to proceed with skin testing or serum allergen-specific IgE testing. What information might have been obtained from such testing?

4. When treating feline asthma, the emphasis is on decreasing inflammation in the lung and improving pulmonary function. How did the drugs that were used in this case accomplish these objectives?

Further Reading

Reinero CR (2011) Advances in the understanding of pathogenesis, and diagnostics and therapeutics for feline allergic asthma. *Vet J* 190:28–33.

Reinero CR, DeClue AE & Rabinowitz P (2009) Asthma in humans and cats: is there a common sensitivity to aeroallegens in shared environments? *Environ Res* 109:634–640.

Trzil JE & Reinero CR (2014) Update on feline asthma. *Vet Clin North Am Small Anim Pract* 44:91–105.

IgE antibodies are a normal defense against parasites that may enter the body of an animal, but sometimes this same response is misdirected towards seemingly harmless material, such as pollens from grass, weeds and trees, food components, dust mites, and mold proteins. When the immune system responds to harmless proteins in the environment by mounting an IgE response, the result is an allergic-type reaction called type I hypersensitivity. Not all animals have a heightened reaction to these substances (called allergens); there is a genetic component that controls the response. Animals that are genetically predisposed to this type of reaction are described as "atopic." The allergy can be manifested in the respiratory tract as hay fever (upper) or asthma (lower), or it may be systemic (anaphylactic shock). When it involves the skin, the resulting syndrome is called atopic dermatitis.

The type I hypersensitivity response requires a sensitization exposure during which the patient either inhales, ingests, or absorbs the allergen through the skin. The allergen is presented by Langerhans cells to T lymphocytes, resulting in their differentiation into T-helper type 2 cells that make cytokines interleukin 4 and interleukin 13. These cytokines stimulate B lymphocytes to differentiate into plasma cells that ultimately make allergen-specific IgE antibodies. Once this process has occurred, the IgE, which has a very high affinity for mast cells that are present within the skin and on mucous membranes, binds to the IgE receptors (FcεRI) on the mast cells (Figure 15.1). The patient is then sensitized, so that subsequent exposure to the same allergen will result in binding of the allergen to the IgE on mast cells and trigger those cells to release a variety of mediators that cause a range of physiological effects. The mast cells contain cytoplasmic granules that themselves contain vasoactive and inflammatory mediators (Table 15.1). In addition to immediate release of these preformed mediators, once the mast cell has

Figure 15.1 Schematic representation (left) and photomicrograph (right) of a mast cell. Mast cells are tissue cells that trigger a local inflammatory response to antigen by releasing substances that act on local blood vessels. They play a key role in atopic dermatitis. (Photograph courtesy of N. Rooney, R. Steinman, and D. Friend. Illustration courtesy of Garland Science.)

Table 15.1 Mast cell mediators that are preformed in mast cell granules are liberated from the cell after antigen cross-links IgE on the cell surface. These mediators have potent physiological effects.

Mast-cell mediators (preformed in granules)	Physiological effects (occurring within minutes)
Histamine	Increases vascular permeability and smooth muscle contraction
Protease, chymase, tryptase	Many, including activation of other cells, and stimulation of CXCL8 (tryptase is a marker for mast-cell activation)
Serotonin	Vasoactive effects; constriction or dilation (species dependent)
Kinins	Vasodilation

TOPICS BEARING ON THIS CASE:

Type I hypersensitivity

Atopy (atopic syndrome)

T-helper type 2 cells and cytokines

Mast-cell mediators

Table 15.2 Stimulation of mast cell membrane bound arachidonic acid metabolism through the lipooxygenase and cyclooxygenase pathways causes production of eicosanoids (leukotrienes and prostaglandins, respectively). These mediators of allergic inflammation take longer to become activated but their effect is potent and longer lasting than that of the pre-formed mediators.

Eicosanoid mediators (formed after mast-cell activation from membrane arachidonic acid)	Physiological effects
Leukotrienes (LT): LTB_4	Attracts neutrophils and stimulates eosinophil chemotaxis
LTC_4, LTD_4, LTE_4	Increase vascular permeability, smooth muscle contraction, and mucus secretion
Prostaglandins (PG): PGE_2	Immune modulation, Th2 cell stimulation
PGD_2	Smooth muscle contraction

been activated by IgE cross-linking (by allergen on its surface), the lipooxygenase and cyclooxygenase enzyme systems are activated to produce lipid mediators from cell membrane arachidonic acid after cleavage from the phospholipid membrane by phospholipase 2. These mediators (known as eicosanoids) have longer-lasting physiological effects than do the preformed mediators (Table 15.2), and in diseases caused by mast-cell degranulation their effects are responsible for many of the clinical signs.

THE CASE OF CODY: A DOG WITH HAIR LOSS AND INFLAMED SKIN WHO COULD NOT STOP SCRATCHING

SIGNALMENT/CASE HISTORY

Cody is an 8-year-old Labrador Retriever cross who has been extremely "itchy" for the past 9 months. He had seen another veterinarian when the clinical signs began, and was put on a tapering course of corticosteroids (prednisone). This helped, and he did not scratch while he was on the medication. However, as soon as the course of therapy was finished he began to scratch again. Now he scratches and bites at his skin incessantly, which has caused significant hair loss and erythematous skin lesions (Figure 15.2). His owner reports that Cody first began licking and biting at his feet, and then progressed to scratching the posterior portion of his body in a variety of sites. He was also described as frequently scratching at his ears. Cody is on heartworm prophylaxis (Heartgard®) and flea prophylaxis (Advantage®). He eats a diet of IAMS® adult dog formula, with occasional treats.

PHYSICAL EXAMINATION

On examination, Cody was bright and alert and in good body condition. His hair coat was sparse over much of his body, with areas of extensive alopecia. There were areas of hyperpigmentation of the skin and lichenification. Erythema was present, as well as evidence of excoriation from self-trauma (scratching). Erythematous lesions on the areas with alopecia appeared to be associated with a mild pyoderma. No fleas or flea dirt were observed. The ears had a slight odor, and otoscopic examination revealed a moderate amount of waxy exudate. All other physical parameters were within normal limits.

DIFFERENTIAL DIAGNOSIS

The chronic nature of the symptoms and the breed of dog suggest that atopic dermatitis is a likely cause of the primary dermatitis. Self-trauma probably

Figure 15.2 Cody showing lesions of atopic dermatitis. Note the areas of alopecia around the eyes and the frontal area of the head. (Courtesy of Danny W. Scott.)

encouraged bacterial infection, causing a secondary pyoderma. Otitis externa also often accompanies dermatologic diseases, such as atopic dermatitis. However, it is necessary to rule out other skin conditions, including flea allergy dermatitis, ectoparasites (*Demodex*, *Sarcoptes*), and food allergy. Hormonal causes of the alopecia are less likely due to the extreme pruritus and the distribution of the lesions.

DIAGNOSTIC TESTS AND RESULTS

A skin scraping failed to show the presence of any mites or dermatophytes. Blood was obtained, and serum was sent to the laboratory for an allergen-specific IgE panel screen. The results showed high reactivity to house-dust mite allergen, and moderate reactivity to mixed weeds and pollens. Intradermal skin testing was performed by injection of a small amount of allergen into the skin. In the allergic patient this process results in local mast-cell degranulation and an erythematous wheal at the site of injection, whereas injection of an allergen to which the patient is not sensitized will not show a wheal. Cody's skin testing showed large wheals in response to the same allergens that were identified by the blood test (Figure 15.3). An ear swab was obtained from each ear and examined under the microscope. Yeast (*Malassezia*) cells were observed. A diet consisting of venison and green peas (commercial prescription diet) was substituted for Cody's normal dog food to evaluate the possibility of food allergy as a cause of or contributing factor to the dermatitis. Food allergens did not appear to be reactive.

DIAGNOSIS

The results of blood and intradermal testing for IgE showed that Cody has a significant allergy to several environmental allergens. He was diagnosed with atopic dermatitis.

TREATMENT

Because it is a chronic condition, treatment of atopic dermatitis is challenging. Avoidance of the allergens is the best solution. However, in most cases this is not possible. The most successful treatments for this condition focus on depression of the immune response—ideally reduction of IgE synthesis. The use of antihistamines has limited success because histamine is only one of many mediators that are responsible for the clinical signs and associated allergic inflammation. Although corticosteroid therapy has value for short-term resolution of clinical signs, continued use of steroids will cause undesirable side effects, and in the worst-case scenario can result in a dog with iatrogenic Cushing's disease.

Once the allergens have been identified as in Cody's case, it is possible to begin a desensitization therapy program (allergy shots). Appropriate allergens are mixed into an injectable solution, which is administered, usually weekly at first, by subcutaneous injection. Gradually the amount of allergen is increased. This treatment is intended to modulate the immune response by increasing regulatory T cells, decreasing Th2 cells, and ultimately shifting antibody production from IgE to IgG. The treatment alternatives presented to Cody's owner included Atopica® a drug that modulates the immune response by suppressing T-cell function and expansion by depression of IL-2 production. Another suggestion was Apoquel®, which stops the itching and inflammation by interfering with the Janus kinase enzyme pathway. Cody's owners elected to try the desensitization therapy, which was quite successful.

ATOPIC DERMATITIS

Canine atopic dermatitis has been defined by the International Task Force on Canine Atopic Dermatitis as a genetically predisposed inflammatory and

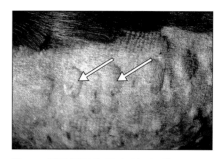

Figure 15.3 Intradermal skin testing showing positive wheals at the site of some allergen injections (arrow), while other injection sites are negative. These positive reactions to injected allergens indicate the presence of mast-cell-bound allergen-specific IgE. (Courtesy of Danny W. Scott.)

pruritic allergic skin disease with characteristic clinical features associated with IgE antibodies most commonly directed against environmental allergens. The task force has also defined the following criteria for diagnosis:

- Onset at under 3 years of age

- Primarily an inside dog

- Pruritus responsive to glucocorticoids

- Initial pruritus occurs prior to the development of lesions

- Front feet and ear pinnae are affected, but not ear margins and the dorso-lumbar area.

There may be additional factors that influence the clinical signs of atopic dermatitis, and these vary between dogs. Some dogs may also have allergies to food antigens that can exacerbate atopic pruritus. Often secondary infections with *Staphylococcus* or *Malassezia* cause accompanying pyoderma.

As shown in Figure 15.4, during acute disease allergens gain entry to the skin, where they are taken up and processed by Langerhans cells. These cells then present the allergen to CD4 T-helper type 1 (Th0) cells, which develop into T-helper type 2 (Th2) cells, making the cytokines IL-4, IL-5, and IL-13.

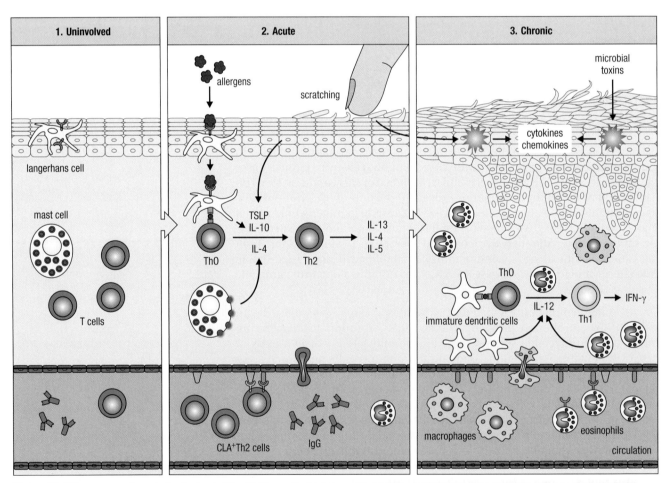

Figure 15.4 The progression of atopic dermatitis. (1) The patient begins with a systemic Th2 cell skew (genetic influence) and IgE. (2) Once allergen is introduced, an acute response ensues with secretion of Th2 cytokines and stimulation of IgE specific for the allergen. Thymic stromal lymphopoietin (TSLP) influences multiple cell types to further facilitate the Th2 response. Degranulation of mast cells further exacerbates the immune skew by secretion of more IL-4. Eosinophils are attracted to the area by the Th2 cytokines and are a characteristic of the disease. (3) Finally, the chronic phase has been postulated (for human patients) to include IL-12 production and a switch to a Th1 cytokine pattern with interferon-γ\ secretion. (Adapted from Leung DY [2000] *J Allergy Clin Immunol* 105:860–876.)

B lymphocytes binding allergen, with second signals from the Th2 cytokines, differentiate into plasma cells that make IgE. The IgE then binds to high-affinity Fc receptors (FcεR1) on mast cells, and upon contact with allergen triggers degranulation of the mast cells. The scratching that occurs is thought to be responsible for much of the tissue damage, and ultimately leads to chronic inflammation with subsequent secondary pyoderma.

COMPARATIVE MEDICINE CONSIDERATIONS

Atopic dermatitis occurs frequently in human patients. Like its canine counterpart, human atopic dermatitis commonly begins in early childhood; approximately 90% of patients show clinical signs before 5 years of age. Patients are characterized as having high serum IgE levels, eosinophilia, expansion of IL4- and IL-13-secreting Th2 cell populations, and a decrease in interferon-γ-producing Th1 cells. In human cases in which histological examination of the skin has been performed, acute lesions are characterized by intercellular edema. Chronic lesional skin contains eosinophils, which undergo cytolysis, liberate cytotoxic granule proteins, and further exacerbate inflammation by producing reactive oxygen intermediates. In studies on children a direct correlation has been found between the presence of food allergy and the severity of atopic dermatitis. The role of aeroallergens has also been shown to increase as children get older. As with canines, human patients with atopic dermatitis have an increased tendency to develop bacterial and fungal skin infections.

The antigens to which a patient is allergic are determined either by intradermal skin testing or by testing serum for specific IgE (in much the same way as in the canine patient). The therapeutic use of desensitization by injection of allergen in small doses over time has not been demonstrated to be effective for treating atopic dermatitis in humans; by contrast, it is a very effective treatment for allergic asthma and rhinitis in many patients.

Atopic dermatitis in cats is quite similar to the disease in dogs. Pruritic cats can present with lesions and clinical signs that are consistent with a diagnosis of atopic dermatitis, but these must be differentiated from other causes of pruritus, such as food and flea allergy or mites. Skin testing of cats is less reliable than in dogs. Allergic/atopic dermatitis in horses is quite common, and is often manifested as urticaria (hives). As in dogs, the causative allergens are determined either by intradermal skin testing or by serum IgE testing. The use of desensitization therapy is successful in slightly over 50% of cases.

Questions

1. Describe the role of the mast cell in atopic dermatitis.

2. What does a serum allergen-specific positive IgE mean? How does this resemble or differ from a positive intradermal skin test for the same allergen? How would you explain the finding of a positive result on the skin test and a negative result on the blood test?

3. Newer drugs that are used to treat atopic dermatitis target immune function more specifically than corticosteroids. Describe the mechanism of action for prednisone, Atopica®, and Apoquel®, respectively.

Further Reading

Leung DY (2000) Atopic dermatitis: new insights and opportunities for therapeutic intervention. *J Allergy Clin Immunol* 105:860–876.

Olivry T, DeBoer DJ, Favrot C et al. (2010) Treatment of canine atopic dermatitis: 2010 clinical practice guidelines from the International Task Force on Canine Atopic Dermatitis. *Vet Dermatol* 21:233–248.

Stepnik CT, Outerbridge CA, White SD & Kass PH (2012) Equine atopic skin disease and response to allergen-specific immunotherapy: a retrospective study at the University of California-Davis (1991–2008). *Vet Dermatol* 23:29–35.

CASE 16
CULICOIDES HYPERSENSITIVITY

Development of an IgE response is an important defense mechanism against parasitic helminths. IgE binds to mast-cell membranes and facilitates a "self-cure" reaction, resulting in expulsion of worms from the intestinal lumen by degranulation with release of histamine and other mast-cell mediators. This is a beneficial function, particularly in a young horse with a heavy burden of roundworms. However, this same mechanism can result in intense pruritus with alopecia when the IgE is directed against salivary antigens of biting flies and gnats. All horses produce IgE antibodies when infested with parasitic roundworms, such as *Ascaris* species, yet only some respond to the bites of flies or gnats by producing IgE in response to antigens injected during the process of blood consumption. This is at least in part because genetics plays an important role in the development of allergy to the bite of an insect; an atopic phenotype has been shown to be inherited in human, canine, and equine cases. This atopic phenotype means that the individual is more likely to mount an IgE antibody response to an antigen than are non-atopic individuals. The genetic predisposition for IgE production in response to environmental allergens has also been associated with polymorphisms in several different immune response genes.

TOPICS BEARING ON THIS CASE:

Type I hypersensitivity

Allergic dermatitis

Genetic predisposition

THE CASE OF CLOWN: A HORSE WITH SEVERE SEASONAL PRURITUS

SIGNALMENT/CASE HISTORY

Clown is a brown Quarter Horse mare that was purchased at 6 years of age in California during the fall months. At the time of purchase her mane was short, but there was no evidence of dermatitis. The previous owner said that she had "rubbed out" her mane and this was why it was short. The following spring Clown developed extremely itchy skin, particularly around the dock of her tail and along the base of her mane (Figure 16.1). She also exhibited hair loss (alopecia) along the inner portion of her pinnae, which became extremely erythematous and sore. The owner of the pasture where Clown was previously kept commented that every spring they get "no-see-ums" that bother the horses at dawn and dusk.

PHYSICAL EXAMINATION

On physical examination Clown was in good body condition, well muscled, and appeared bright and alert. She strongly resisted having her ears handled. Hair within the pinnae was sparse, and erythematous crusty lesions were apparent. Examination of her neck revealed erythematous lesions with areas of excoriation and alopecia along the dorsum of the neck and along the dock of the tail. No additional skin lesions were observed.

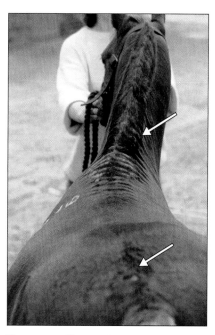

Figure 16.1 Clown exhibiting symptoms of dermatitis with severe excoriation and alopecia along the dorsal neck and tail dock (arrows). (Courtesy of Derek Knottenbelt.)

DIFFERENTIAL DIAGNOSIS

Culicoides hypersensitivity was the most likely diagnosis based on history and clinical signs. However, atopic dermatitis, food allergy, ectoparasite infestation, and non-*Culicoides* insect hypersensitivity were also considered. Allergy to feed can manifest as skin lesions, but the location on the body is usually more generalized. Infestation with lice or ticks would not cause the type of lesions in the specific locations that were present on Clown. The lice or ticks themselves can be identified on the animal, and the location is generally less centralized to specific areas. Lesions seen as a result of hypersensitivity are primarily caused by self-trauma due to the intense itching that occurs after the mast-cell mediators are liberated into the tissue.

DIAGNOSTIC TESTS AND RESULTS

To rule out ectoparasite infestation, a bright light and flea comb were used to determine whether lice were present. Tape preparations and skin scrapings were used to detect the presence of chorioptic and psoroptic mange mites, *Trombicula* larvae, and forage and poultry mites. None of these organisms were observed. Eggs of *Oxyuris* species were not found. A fungal culture performed on hair for dermatophytes was negative. A tentative diagnosis of *Culicoides* hypersensitivity was made, and blood (red-top tube for serum) was submitted for allergen-specific IgE testing for *Culicoides* and other flies. The test was positive for *Culicoides* IgE (moderately high level). Another test that could have been performed is intradermal skin testing with allergens. However, this type of testing is more invasive (it requires sedation, clipping, and administration of numerous injections), and is considered by many to be less reliable in the horse than in humans and dogs.

DIAGNOSIS

Based on the serum test results for *Culicoides* IgE, Clown was diagnosed with *Culicoides* hypersensitivity, also known as summer eczema or "sweet itch."

TREATMENT

Clown's owner was advised to stable her during the morning, turn her out in the afternoon, and return her to the barn before dusk. Clown was treated with a short course of prednisolone (1 mg/kg/day for 7 days, then 0.5 mg/kg for a further 7 days) to dampen the inflammatory response. Her ears were treated with a dexamethasone antibiotic ointment, and a fly mask with ear covers was prescribed. Clown was also fitted with a permethrin-impregnated fly sheet that covered her belly, neck, and dorsal tail. Her owner was advised that immunotherapy was an option, but she decided to evaluate the protocol for fly exposure reduction before taking this step. Some studies have shown good improvement with allergen-specific immunotherapy, while others have not. Future use of recombinant *Culicoides* allergens may increase the usefulness of this therapy.

CULICOIDES HYPERSENSITIVITY

The *Culicoides* gnat is a very small flying insect (1–3 mm long) belonging to the Ceratopogonidae family (Figure 16.2). It is common in temperate climates, and its appearance is seasonal, from spring to autumn. The female takes blood meals, primarily in the early morning and evening. Preferential feeding sites include the mane, tail, ears, and ventral midline of the horse. Development of allergic dermatitis in response to the bite of these gnats is more common in horses that reside in open pasture than in those that are stabled. The allergy can occur in any breed of horse, but is more common in Icelandic horses that are exported as young adults (the insect does not exist in Iceland), Shetland ponies, Swiss warmbloods, Friesians, and Quarter Horses. Increased frequency of the

Figure 16.2 The *Culicoides* gnat. (Adapted from Wikimedia Commons.)

allergy in horses with certain equine leukocyte antigen (ELA) specificities suggests that genetics may play a role. Horses of any age can be affected, but the dermatitis often starts by 3–4 years of age and progressively worsens each season. The bite of the gnat causes extreme pruritus, which results in excoriation from self-trauma (rubbing) to alleviate the itch. The distribution of the lesions may be primarily dorsal (as in Clown's case) or ventral, or less commonly dorsal and ventral. The pathogenesis of the disease involves development of IgE antibodies to one or more antigens within the saliva of the gnat. At least 11 antigens have been identified as having elicited positive intradermal tests in horses with the allergy. Immune studies performed to date suggest that there is a strong T-helper type 2 response, and possibly a deficiency of regulatory T cells in infected horses (which facilitates the IgE response). Mast-cell degranulation occurs with successive gnat bites (Figure 16.3). Histopathology of the lesions shows evidence of a type I hypersensitivity response, with perivascular infiltration of eosinophils. Mast cells (tryptase positive) are present in increased numbers, and some studies have shown CD4$^+$ T cells infiltrating the dermis.

COMPARATIVE MEDICINE CONSIDERATIONS

Culicoides gnats are somewhat host specific in their feeding habits, preferring equine and ruminant hosts to humans and small carnivores. Although feeding of *Culicoides* on bovine and ovine hosts is the mode of infection for the bluetongue virus, the development of allergy to the salivary antigens of this gnat is not a documented problem in these species. Thus the *Culicoides* hypersensitivity syndrome is unique to equines. Type I hypersensitivity to other insect bites has been described in many species—for example, allergic dermatitis in dogs and cats (flea allergy dermatitis; see Case 28), and hypersensitivity to mosquito bites in humans.

Questions

1. Why is the dermatitis caused in the horse by *Culicoides* species localized to certain areas on the horse's body?

2. Describe the immune response that results in sensitization and elicitation of the IgE response to the bite of the *Culicoides* gnat. In what respects is it similar to or different from the development of allergic asthma in the cat (see Case 14)?

3. Intradermal testing with salivary antigens from the gnat is sometimes positive for horses that are not clinically affected with seasonal *Culicoides* hypersensitivity dermatitis. How might you explain this?

Further Reading

Marti E & Hamza E (2014) Equine immunoglobulin E. In Veterinary Allergy (Noli C, Foster A & Rosenkrantz W eds.), pp 279–286. Wiley Blackwell.

Wagner B (2014) Pathogenesis and epidemiology of *Culicoides* hypersensitivity. In Veterinary Allergy (Noli C, Foster A & Rosenkrantz W eds.), pp. 275–278. Wiley Blackwell.

Wilson AD (2014) Immune responses to ectoparasites of horses, with a focus on insect bite hypersensitivity. *Parasite Immunol* 36:560–572.

	Type I
Immune reactant	IgE
Antigen	soluble antigen
Effector mechanism	mast-cell activation
	Ag
Example of hypersensitivity reaction	allergic rhinitis, allergic asthma, atopic eczema, systemic anaphylaxis, some drug allergies

Figure 16.3 The mechanisms involved in a type I hypersensitivity response. In *Culicoides* hypersensitivity, IgE antibodies produced in response to salivary antigens of the *Culicoides* gnat bind to Fc epsilon receptors on dermal mast cells. Repeated introduction of the antigens during successive bites activates mast cells to degranulate by IgE cross-linking. Mediators such as histamine cause intense pruritus, which stimulates rubbing and excoriation. Eosinophils are attracted to the area, and increased vascular permeability causes tissue edema. (From Murphy K [2011] Janeway's Immunobiology, 8th ed. Garland Science.)

CASE 17
SYSTEMIC ANAPHYLAXIS

The immune response is intended to be protective. However, sometimes antigens that are not actually a threat to the health of the animal elicit the production of IgE antibodies, as when inhaled pollens cause sneezing in people with hay fever. This is called a type I hypersensitivity response. The IgE antibodies are present in very low amounts (of the order of nanograms) in the circulation, but have a very strong affinity for receptors on mast cells in tissues, and can remain on those cells for weeks to months. Mast cells are located in the connective tissue of the body and at mucosal sites. Because of the high-affinity receptors for IgE (FcεRI) on their membranes, they are able to bind the IgE as it is released from plasma cells. When a patient becomes sensitized to an antigen, IgE is produced. The IgE binds to the mast cells, and the next time the patient is exposed to that antigen an allergic response occurs. The allergic response is initiated when the IgE attaches to the mast cell by its Fc end, binds the antigen (called an allergen) by its Fab portions, and there is cross-linking of the IgE on the surface of the mast cells. This cross-linking induces degranulation of the cell, releasing many vasoactive mediators. Preformed mediators that are released include histamine (which causes vasodilation, smooth muscle contraction, and increased vascular permeability), proteases, proteoglycans, and some cytokines. Synthesis of several mediators is also initiated, including the arachidonic acid metabolites, namely prostaglandins and leukotrienes. These mediators collectively cause prolonged smooth muscle contraction and inflammatory cell chemotaxis (Figure 17.1). When this type of reaction is localized to the skin, a wheal forms due to the increase in local

TOPICS BEARING ON THIS CASE:

IgE-mediated mast-cell degranulation

Environmental sensitization to allergens

Type I hypersensitivity

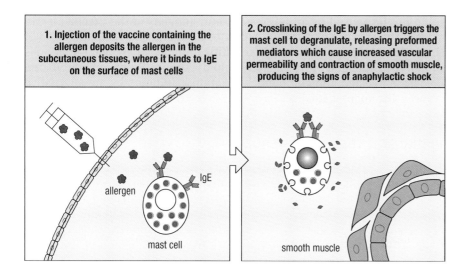

1. Injection of the vaccine containing the allergen deposits the allergen in the subcutaneous tissues, where it binds to IgE on the surface of mast cells

2. Crosslinking of the IgE by allergen triggers the mast cell to degranulate, releasing preformed mediators which cause increased vascular permeability and contraction of smooth muscle, producing the signs of anaphylactic shock

allergen

IgE

mast cell

smooth muscle

Figure 17.1 1. Injection of the vaccine containing the allergen deposits the allergen in the subcutaneous tissues, where it binds to IgE on the surface of mast cells. 2. Cross-linking of the IgE by allergen triggers the mast cell to degranulate, releasing preformed mediators that produce the physiological effects that cause anaphylactic shock.

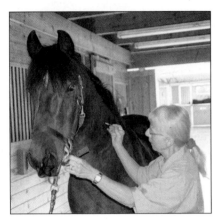

Figure 17.2 Jet receiving his vaccination in the lateral neck muscle. Just minutes later he exhibited dyspnea and signs of shock.

fluid release resulting from enhanced vascular permeability. The intradermal skin testing that is used to diagnose allergy is an example of such a localized reaction (see Case 15). However, when this response occurs throughout the patient, the result can be fatal shock due to hypotension and/or inability to breathe (resulting from smooth muscle contraction of bronchi), a phenomenon known as systemic anaphylaxis

THE CASE OF JET: A HORSE WHO HAD A LIFE-THREATENING REACTION TO A VACCINE

SIGNALMENT/CASE HISTORY

Jet is a 15-year-old bay Morgan gelding with a history of good health. He has been vaccinated annually since he was a yearling for eastern and western equine encephalitis virus, rabies, equine influenza virus, tetanus, and equine herpes virus. His owner recently moved from a chicken farm, where Jet had lived his entire life, to the city, and Jet is now stabled at an equestrian center. Jet's problem began with a vet visit to his new barn. He received his annual vaccines along with 20 other horses that were boarded at the barn. The vaccines were administered intramuscularly into the gluteal and neck muscles (Figure 17.2). As the veterinarian was packing up to leave the barn she was summoned to return quickly because Jet was acting strangely.

PHYSICAL EXAMINATION

On physical examination, Jet was sweating, breathing rapidly, and anxious, and he appeared close to collapse. On examination his capillary refill was slow; he was dyspneic and passed some watery diarrhea.

DIFFERENTIAL DIAGNOSIS

The history of good health immediately prior to receiving the vaccination narrows the differential diagnosis to something associated with the administration of the vaccine. The vet had aspirated slightly before injecting the vaccine into the muscle, ruling out an accidental intravenous or intraarterial injection. The vaccines were from the same lot as those administered without incident to the other 20 horses in the barn, thus ruling out the possibility that there was some toxic component in the vaccine. A reaction that resembles anaphylactic shock, called an anaphylactoid reaction, sometimes occurs in horses and other animals. An anaphylactoid reaction would not be caused by an IgE response to antigens in the vaccine; it can occur in the absence of previous exposure to an antigen. Contamination of a biologic with lipopolysaccharide resulting from bacterial contamination is sometimes associated with an anaphylactoid response. When a contaminated biologic is injected and complement fixation is initiated by the alternative pathway, the anaphylatoxins C3a and C5a are produced and cause mast-cell degranulation.

DIAGNOSTIC TESTS AND RESULTS

The presence of IgE against one or more vaccine components as demonstrated by a serum or skin test can differentiate between an anaphylactoid response and an anaphylactic reaction. However, in cases of immediate system distress such as this one, there is no time to perform such a test. If the reaction is treated and the horse survives, it is important to determine the causal antigen using one of these tests before any vaccines are given in the future.

DIAGNOSIS

A tentative diagnosis, of a type I hypersensitivity response to something in the vaccine, was made. Upon further questioning of the owner, it was discovered that Jet had developed a large wheal at the vaccine site the previous year, providing evidence for a diagnosis of systemic anaphylaxis in response to the vaccine(s).

TREATMENT

An intravenous bolus of epinephrine and an intramuscular injection of dexamethasone were administered immediately. This was followed by another epinephrine injection 15 minutes later. Jet was closely monitored over the next few hours. He was treated with intramuscular dexamethasone for the following 2 days, and made a full recovery.

The owner was advised that Jet was very probably allergic to something in one of the vaccines, and that if the veterinarian had not still been on the premises when Jet began to develop clinical signs, it is very unlikely that he would have survived the anaphylaxis. The swelling at the vaccine site after vaccination the previous year was a sign that allergic reactivity was developing. Since Jet is now 15 years old, and many veterinary immunologists are advocating longer time periods between vaccinations, the owners were advised not to re-vaccinate Jet for at least 3 years, and possibly longer.

SYSTEMIC ANAPHYLAXIS

Systemic anaphylaxis, or anaphylactic shock, is a true emergency that requires the immediate use of epinephrine to counteract the smooth muscle contraction produced by mast-cell mediators. Vaccine reactions occur infrequently in horses, dogs, and cats. However, the severe nature of the response means that it is important to be aware of animals that are at risk. Recent studies on horses and dogs have shown that it is often not the viral antigens themselves that elicit the IgE response, but rather proteins that appear in the vaccine as contaminants. For example, virus that is grown in tissue culture with fetal bovine serum in the media is harvested for vaccine use and not completely purified. The remaining fetal bovine serum proteins can stimulate an IgE response in the vaccinated patient. Since these contaminants can be present in several different vaccines, the response that is stimulated by a rabies vaccine, for example, could be elicited by a West Nile virus vaccine. In this case, it is highly likely that the historical exposure of Jet to chickens had caused sensitization to avian proteins, which were present in the influenza vaccine that Jet received.

As discussed in Case 15, animals and humans who respond to antigen by producing IgE are described as "atopic" because they have a genetic predisposition to the allergic response. Anaphylactic shock most commonly occurs when antigen is introduced by injection, although in extremely sensitive patients other modes of introduction, such as inhalation or oral ingestion, may also cause anaphylaxis (this is particularly true of human patients with shellfish or peanut allergy). The clinical signs vary according to the shock organ(s) involved. Species differences in shock organs and the major mediators produced are shown in Table 17.1. The low occurrence of type I hypersensitivity reactions to vaccines in horses is probably due to equine genetic composition and its influence on IgE responsiveness. Only a small percentage of the equine population develops allergy and is considered atopic; it is this population that preferentially produces IgE in response to the contaminating foreign proteins present in vaccines. The majority of horses will respond immunologically by mounting an IgG response to these non-target antigens—a response that does not cause anaphylaxis.

Table 17.1 Systemic anaphylaxis in different species. (Adapted from Tizard IR [2008] Veterinary Immunology: An Introduction, 8th ed. Saunders.)

Species	Shock organs	Clinical signs	Pathology	Major mediators*
Horse	Respiratory tract; intestine	Cough Dyspnea Diarrhea	Pulmonary emphysema Intestinal hemorrhage	Histamine Serotonin
Cattle, sheep, goats	Respiratory tract	Cough Dyspnea Collapse	Pulmonary edema Emphysema Hemorrhage	Histamine Serotonin Leukotrienes
Swine	Respiratory tract; intestine	Cyanosis Pruritus	Systemic hypotension	Histamine
Dog	Hepatic veins	Dyspnea Diarrhea Vomiting Collapse	Hepatic congestion Visceral hemorrhage	Histamine Leukotrienes Prostaglandins
Cat	Respiratory tract; intestine	Dyspnea Diarrhea Vomiting Pruritus	Lung edema Intestinal edema	Histamine Serotonin Leukotrienes
Human	Respiratory tract	Dyspnea Urticaria Collapse	Lung edema Emphysema	Histamine Leukotrienes

*Mediator shows some of the species differences.

COMPARATIVE MEDICINE CONSIDERATIONS

IgE responses to non-target antigens in vaccines have been described in human patients for several different vaccines. Hypersensitivity responses to human vaccines have also been reported for several vaccine components. Gelatin, which is often included as a stabilizer in vaccines, has been implicated in anaphylaxis associated with measles, mumps, and rubella (MMR) vaccines. Sensitization to gelatin has also been stimulated by diphtheria, tetanus, and acellular pertussis (DPT) vaccination in children. Human patients who are allergic to eggs have undergone anaphylaxis after being immunized with live attenuated influenza vaccine grown in eggs. Studies in dogs reacting to viral vaccines have shown similar specificity to that seen both in horses (fetal bovine serum proteins) and in humans (gelatin).

Questions

1. Describe how a non-target vaccine antigen sensitized and elicited an allergic reaction in this case.

2. Of what is the wheal that developed at the injection site after vaccination the previous year symptomatic? Is there a connection between the appearance of this wheal and the vaccine that Jet recently received?

3. How does epinephrine work to "save" a patient with anaphylaxis?

Further Reading

Gershwin LJ, Netherwood KA, Norris MS et al. (2012) Equine IgE responses to non-viral vaccine components. *Vaccine* 30:7615–7620 (doi: 10.1016/j.vaccine.2012.10.029).

Hernandez L, Papalia S & Pujalte GG (2016) Anaphylaxis. *Prim Care* 43:477–485.

Nakayama T, Aizawa C & Kuno-Sakai H (1999) A clinical analysis of gelatin allergy and determination of its causal relationship to the previous administration of gelatin-containing acellular pertussis vaccine combined with diphtheria and tetanus toxoids. *J Allergy Clin Immunol* 103:321–325.

Oettgen HC (2016) Fifty years later: Emerging functions of IgE antibodies in host defense, immune regulation, and allergic diseases. *J Allergy Clin Immunol* 137:1631–1645.

Ohmori K, Masuda K, Maeda S et al. (2005) IgE reactivity to vaccine components in dogs that developed immediate-type allergic reactions after vaccination. *Vet Immunol Immunopathol* 104:249–256.

CASE 18
NEONATAL ISOERYTHROLYSIS

EQUINE

Equine erythrocytes (red blood cells) express antigens that constitute their blood type; these antigens are inherited as dominant alleles, and the frequency with which these dominant genes are expressed differs across horse populations. Some of the erythrocyte antigens are strongly immunogenic for horses that do not share the same antigens on their own red blood cells. In contrast to the human erythrocyte ABO system, in which an individual who lacks an antigen always produces antibodies against that antigen (for example, a person with blood type A has antibodies to the B antigen), the horse does not automatically have antibodies against an erythrocyte antigen that it lacks. An exception occurs when a brood mare is repeatedly bred to a stallion which has erythrocyte antigens that the mare's cells lack. During parturition, there is an opportunity for some of the fetal blood to enter the maternal circulation, thus allowing the mare to mount an immune response to the erythrocyte antigens present on the foal's erythrocytes. This does not usually affect the current foal, but the mare's next foal could be at risk if the sire has transmitted the same dominant erythrocyte antigen. This subsequent foal is born bright and alert and begins to suckle soon after standing. The lack of transplacental transfer of immunoglobulin in the equine protects the foal *in utero*. However, once the foal suckles colostrum, the antibodies that the dam has made against the dominant antigens on the foal's erythrocytes are adsorbed, and within 72 hours the foal becomes lethargic, dyspneic, and shows signs of anemia.

Destruction of the foal's erythrocytes occurs as the result of a type II hypersensitivity reaction. There are three ways in which antibodies specific for erythrocyte antigens can destroy these erythrocytes. First, antibodies can bind to the cell, fix complement, and lyse the cell using the classical complement pathway. Second, antibodies can opsonize the erythrocytes for removal in the spleen by fixed phagocytes. Third, C3b (produced by complement fixation) can bind to C3b receptors on erythrocytes and opsonize the cells for removal in the spleen. In each case the result is loss of erythrocytes, and anemia (Figure 18.1).

THE CASE OF IZZIE: A COLT WHO COLLAPSED AND WAS IN SEVERE DISTRESS AFTER SUCKLING COLOSTRUM

SIGNALMENT/CASE HISTORY

Izzie is a Thoroughbred-Morgan cross colt (his dam was a Morgan and his sire was a Thoroughbred). The mare had been bred to the same stallion and produced a foal from this mating the previous 2 years. Izzie appeared to be a normal healthy foal at birth. He suckled within a few hours and was bright and

TOPICS BEARING ON THIS CASE:

Equine blood types

Coombs' test

Type II hypersensitivity

Complement fixation

Passive antibody transfer

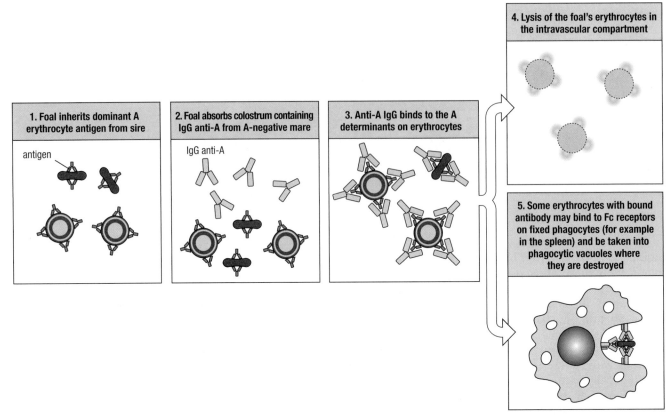

Figure 18.1 (1) The foal has inherited dominant A erythrocyte antigen from the sire. (2) The mare (dam) is A negative and has produced anti-A IgG antibodies as a result of exposure to A erythrocytes in previous pregnancies. These are present in her colostrum. (3) After consuming the colostrum, the foal absorbs the antibodies into its circulation and the anti-A IgG binds to the A determinants on the foal's erythrocytes. (4) Complement is fixed by the classical pathway, resulting in lysis of the foal's erythrocytes in the intravascular compartment. (5) Some erythrocytes with bound antibody may bind to Fc receptors on fixed phagocytes (for example, in the spleen) and be taken into phagocytic vacuoles, where they are destroyed.

responsive. The day before presentation to the veterinarian he seemed to be depressed and was not nursing well. His condition worsened and the veterinarian was called. Upon arrival the vet noted that Izzie refused to rise, and his oral mucous membranes and sclera were yellow in color, indicating possible icterus. The colt was sent to the veterinary hospital for further diagnostics and treatment.

PHYSICAL EXAMINATION

At presentation, Izzie was weak and depressed with a body temperature of 99.9°F (normal range, 100–102°F), pulse of 92 beats/minute (normal range, 60–80 beats/minute), and a respiratory rate of 28 breaths/minute (normal range, 20–40 breaths/minute). Izzie was a normal term foal and he did not show evidence of trauma or physical defects. His joints were not enlarged, and gastrointestinal motility was normal. His mucous membranes and sclera were pale yellow (icterus) (Figure 18.2), and capillary refill was poor. A hematocrit obtained at presentation was 16%.

DIFFERENTIAL DIAGNOSIS

Neonatal isoerythrolysis (NI) is the most likely diagnosis for a foal that is born healthy but deteriorates within 3 days to a near-death condition with a low hematocrit. Other conditions that can cause rapid decline in neonatal foals include septicemia and congenital cardiac defects. The icterus and low hematocrit with normal joints and no evidence of diarrhea support the diagnosis of NI.

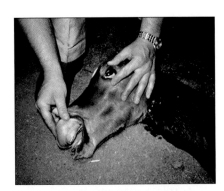

Figure 18.2 Izzie displaying yellow-colored mucous membranes. (Courtesy of Maarten Drost, University of Florida, College of Veterinary Medicine.)

Table 18.1 Abnormal results identified in equine chemistry panel

Parameter	Measured value	Horse reference range
Anion gap (mmol/L)	17	3–9
Alkaline phosphatase (IU/L)	585	86–285
Gamma-glutamyl transferase (GGT) (IU/L)	32	8–22
Total bilirubin (IU/L)	16.6	0.5–2.3
Direct bilirubin (IU/L)	2.4	0.2–0.6

DIAGNOSTIC TESTS AND RESULTS

An equine chemistry panel (Table 18.1) and equine hematology panel (Table 18.2) were performed. The hematology panel revealed that Izzy was very anemic, with a hematocrit of 16% (normal range, 30–46%), red blood cell count of 3.95×10^6 (normal range, $8.2–12.3 \times 10^6$), and a hemoglobin concentration of 5.8 g/dL (normal range, 11.5–17.1 g/dL). Leukocyte counts were within normal limits. The chemistry panel showed that Izzie was icteric, with a total bilirubin concentration of 16.6 IU/L (normal range, 0.5–2.3 IU/L) and a direct bilirubin concentration of 2.4 IU/L (normal range, 0.2–0.6 IU/L). The icterus was a result of massive intravascular hemolysis. The anemia had resulted in an anion gap. A SNAP® test was performed to determine the approximate amount of maternal IgG that Izzie had absorbed. This value was found to be 800 mg/dL (which is within the normal range). A direct Coombs' test for antibody bound to erythrocytes was performed and was positive. Direct slide autoagglutination was also positive. These two tests confirm that that there were antibodies attached to Izzie's erythrocytes.

DIAGNOSIS

A diagnosis of neonatal isoerythrolysis (NI) was made based on anemia, icterus, and the positive Coombs' and autoagglutination tests.

TREATMENT

Izzie was treated with nasal oxygen insufflation while blood was taken from his dam for transfusion. The erythrocytes were washed twice to remove antibodies, and the cells were then resuspended in physiological saline solution for transfusion to Izzie. The foal was also treated with dexamethasone and glucose, as his blood glucose concentration was low at presentation. After he had been stabilized, a nasogastric tube was placed and he was fed mare's milk plus Foal-Lac®, a commercial dehydrated mare's milk replacement formulation, as well as goat's milk. Izzie gradually improved over the next few days, and was ultimately able to nurse on his own. He recovered and was returned home to his owner.

Table 18.2 Equine hematology panel results

Erythrocyte parameter	Measured value	Foal reference range
Red blood cell count $\times 10^6$	3.95	8.2–12.3
Hemoglobin (g/dL)	5.8	11.5–17.1
Hematocrit (%)	16.0	30–46
Leukocyte parameter	**Measured value**	**Foal reference range**
Bands	2640	Rare
Monocytes/µL	0	50–550
Other		
Fibrinogen (mg/dL)	20	100–300
Icterus index	75	> 15 indicates jaundice

Table 18.3 Population frequencies for horses that lack factors Aa or Qa

Breed	Aa-negative	Qa-negative
Thoroughbred	0.02	0.15
Arabian	0.03	0.63
Standardbred	0.19	0.99
Morgan	0.19	0.99
Quarter Horse	0.26	0.68
Peruvian Paso	0.22	0.96

Testing of Izzie's dam showed that she was Qa negative, the stallion was Qa positive, and thus Izzie was also Qa positive. The antibodies in the mare's blood were therefore of the anti-Qa specificity. The owner was instructed that before choosing a stallion if the mare was bred again, her blood should be tested for antibodies reactive with equine erythrocyte antigens. If this precaution was not taken and a pregnancy occurred without previous testing, the foal should not be allowed to suckle the mare after birth, and an alternative supply of colostrum would need to be provided.

NEONATAL ISOERYTHROLYSIS

The equine blood type antigens most often associated with neonatal isoerythrolysis are Aa and Qa. Mares lacking these erythrocyte antigens that are bred to stallions expressing these antigens are most likely to produce antibodies that bind to the antigens on the foal's erythrocytes. Not all Qa- or Aa-negative mares that give birth to Qa- or Aa-positive foals develop antibodies to these antigens; it is not known why some do and others do not. However, once a mare has given birth to a foal with NI, all subsequent foals are at risk. There are some breed frequency differences that can be taken into consideration when trying to find an appropriate stallion to mate with a mare with a known NI foal history, as shown in Table 18.3. NI antibody screening can be performed to test for the presence of antibodies in mare serum that react with specific equine erythrocyte antigens. This utilizes both lytic and agglutination testing and a panel of erythrocytes from horses with known blood types.

Izzie's treatment involved transfusion with his dam's erythrocytes. The mare's erythrocytes are the best cells to use because they lack the Qa antigen, for which Izzy now has circulating antibodies. The plasma will contain the pathogenic antibodies, so it is important to wash the dam's erythrocytes free of antibodies and give only the cells. This procedure is easily performed by centrifugation and then resuspension of erythrocytes in physiological saline solution after the plasma has been decanted. The mare's blood is readily available and is thus a convenient source if there are nearby laboratory facilities for washing the cells. If a gelding with the appropriate blood type is available, whole blood transfusion is also an option.

COMPARATIVE MEDICINE CONSIDERATIONS

Rh disease (erythroblastosis fetalis) in humans is similar to NI in equines, although the antigens involved and the timing of disease induction are different. The human infant that suffers from Rh disease becomes affected *in utero*, whereas the equine foal is not affected until after birth. This disparity reflects a difference in placentation. In the human, the placenta has two layers between the maternal and fetal circulation, and IgG is transported across from the maternal to the fetal circulation during pregnancy. In contrast, the equine placenta has six layers, and there is no transplacental IgG transfer; instead, colostral antibodies consumed by nursing are transported from the gut lumen to the fetal circulation. The mechanism of anemia is similar in both species—antibody and complement opsonize and/or lyse the erythrocytes, removing them from the circulation and creating an icteric and anemic fetus or foal (a classic type II hypersensitivity reaction). Women are routinely tested to determine whether they have the Rh antigen (Rh positive); if they do, there is no risk of their having an Rh baby. However, if a woman is Rh negative and has a child with an Rh-positive man, an Rh-diseased baby could result if the woman has multiple pregnancies. The incidence of Rh disease in women has decreased since the introduction of a therapeutic antibody (RhoGAM®). Immediately after giving birth to an Rh-positive baby, the Rh-negative mother is injected with this anti-Rh antibody, which effectively prevents any fetal erythrocytes that might have entered the mother's circulation from sensitizing her immune system to the foreign antigen.

NI has also been reported in kittens, when animals with blood type A or AB receive colostral anti-A alloantibodies from a type B queen. Such kittens must be removed from the queen on the first day of life, and can safely be fed milk from a type A queen.

Questions

1. Explain how consumption of colostrum causes erythrocyte destruction in a foal with NI.

2. Why would you not expect a first-foal mare to give birth to a foal that develops NI?

3. Izzy had a high icterus index, which was not surprising as his mucous membranes were yellow in color. What caused the icterus and the yellow coloration of the mucosa?

Further Reading

Bailey E (1982) Prevalence of anti-red blood cell antibodies in the serum and colostrum of mares and its relationship to neonatal isoerythrolysis. *Am J Vet Res* 43:1917–1921.

Boyle AG, Magdesian KG & Ruby RE (2005) Neonatal isoerythrolysis in horse foals and a mule foal: 18 cases (1988–2003). *J Am Vet Med Assoc* 227:1276–1283.

Giger U & Casal ML (1997) Feline colostrum—friend or foe: maternal antibodies in queens and kittens. *J Reprod Fertil Suppl* 51:313–316.

CASE 19
IMMUNE-MEDIATED
HEMOLYTIC ANEMIA

The development of antibodies against self cells and tissues is an abnormal occurrence that implies a loss of self tolerance. Despite the purging of self-reactive T cells that occurs in the thymus during fetal life, T cells capable of responding to self antigens by induction of an immune response can remain present in the secondary lymphoid tissues. Some B cells that are reactive to self can also survive during B-cell maturation in the bone marrow. This is because self tolerance of B cells for some antigens is only maintained because there is an absence of T-cell help. The introduction of cross-reacting antigens for which responsive T cells exist may be all that is needed to activate self-reactive B cells. These B cells can then bind to self antigen with their B-cell receptors and receive co-stimulatory signals from T cells and cytokine signals—the three signals necessary to induce the B cells to proliferate, differentiate, and ultimately form plasma cells that produce antibody against the self antigen.

In immune-mediated hemolytic anemia (IMHA), the production of antibodies against self erythrocytes leads to erythrocyte destruction and resultant anemia. The antibody isotype will dictate which immune destructive mechanism predominates. However, destruction always occurs via a type II response initiated by either IgG or IgM binding, as shown in Figure 19.1.

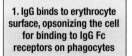

TOPICS BEARING ON THIS CASE:

Autoimmunity

Type II hypersensitivity

Loss of self tolerance

| 1. IgG binds to erythrocyte surface, opsonizing the cell for binding to IgG Fc receptors on phagocytes | 2. IgG binds to IgG Fc receptors on phagocytes, such as fixed macrophages in the spleen | 3. These cells engulf the opsonized erythrocytes and destroy them in phagocytic vacuoles | 4. IgM autoantibodies generally lyse erythrocytes by fixing complement through the classical pathway | 5. IgG antibodies can also lyse erythrocytes when a doublet forms and fixes C1q on the cell surface to initiate the complement cascade |

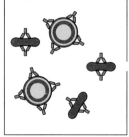 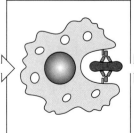 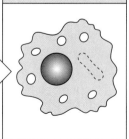 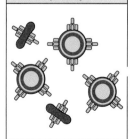 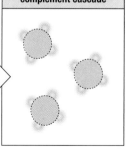

Figure 19.1 In immune-mediated anemia, antibodies are made against self-antigenic determinants on erythrocytes. These can be IgG and/or IgM. (1) IgG bound to the erythrocyte surface opsonizes the cell for binding to IgG Fc receptors on phagocytes (2), such as fixed macrophages in the spleen. (3) These cells engulf the opsonized erythrocytes and destroy them in phagocytic vacuoles. (4) IgM autoantibodies generally lyse erythrocytes by fixing complement through the classical pathway. (5) IgG antibodies can also lyse erythrocytes when a doublet forms and fixes C1q on the cell surface to initiate the complement cascade.

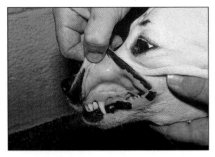

Figure 19.2 Brody's mucous membranes were white, indicating severe anemia. (From Schaer M [2009] Clinical Medicine of the Dog and Cat, 2nd ed. Courtesy of CRC Press.)

THE CASE OF BRODY: A DOG WITH ACUTE ONSET OF WEAKNESS, INAPPETENCE, AND VOMITING

SIGNALMENT/CASE HISTORY

Brody is a 5-year-old castrated male mixed-breed dog. He presented to the emergency service of the veterinary hospital in some distress. He was unwilling to walk or stand, and there was evidence of vomit on the fur around his mouth. His owner reported that Brody had been well up until a few days ago, when he became lethargic, refused to eat his food, and vomited some yellow fluid several times. Brody had a good history of health prior to the present episode, and was not on any chronic medication. He is an indoor dog who has supervised walks and no history of trauma or travel to woody areas. There was no apparent exposure to toxins within the home or garden.

PHYSICAL EXAMINATION

On physical examination, Brody's temperature was normal at 101.3°F (normal range, 100.5–102°F), his pulse was high at 170 beats/minute (normal range, 60–140 beats/minute) with tachycardia, and he was panting. His mucous membranes were white (Figure 19.2), and he was not willing to stand for examination. His hair coat was clean and full, with no evidence of ectoparasites. On palpation his abdomen was normal, and on auscultation the respiratory tract did not reveal any wheezes or crackles. Peripheral lymph nodes were within normal limits. Examination of the neurological system did not reveal any abnormalities.

DIFFERENTIAL DIAGNOSIS

The presence of very pale mucous membranes accompanied by severe lethargy was an indicator of anemia. There was no evidence of bleeding (petechiae, ecchymoses, etc.), so the likelihood that thrombocytopenia was a primary diagnosis was low. Causes of anemia include decreased erythrocyte production, increased erythrocyte destruction (possibly immune mediated), and loss due to hemorrhage. Other causes include chronic infectious or inflammatory conditions.

DIAGNOSTIC TESTS AND RESULTS

Initial blood work included a complete blood count (CBC) and chemistry panel. Other tests are usually performed if the results of the CBC indicate severe anemia (as in this case). These included a cross-match for a potential transfusion, and a Coombs' test to determine whether immune-mediated mechanisms were potentially involved in the apparent anemia. An abdominal ultrasound examination was performed to evaluate splenic and hepatic size, and a coagulation panel was also performed.

On the CBC the red blood cell count was $1.07 \times 10^6/\mu L$ (normal range, 5.6–8.0 $\times 10^6/\mu L$), the hemoglobin concentration was 3.8 g/dL (normal range, 14–19 g/dL), and the hematocrit was 8.4% (normal range, 40–55%). Occasional spherocytosis was noted. The white blood cell count was elevated at 36,930/μL (normal range, 6000–13,000 /μL), with an increase in the absolute neutrophil count, which was 28,914/μL (normal range, 3000–10,500/μL). There were increased band neutrophils and also monocytes. Platelet levels were within the normal range. The biochemistry panel revealed an increase in the total bilirubin concentration to 1.5 mg/dL (normal range, 0–0.2 mg/dL), and an elevated fibrinogen concentration at 624 mg/dL. The direct Coombs' test was performed to evaluate the presence of autoantibodies bound to erythrocytes, and was positive at 1:128. The abdominal ultrasound examination showed moderate splenomegaly without nodules or masses. To prepare for the possible need for transfusion, a cross-match was performed with 2+ agglutination with all potential donors, including self (autoagglutination).

The results of the CBC confirmed that Brody was severely anemic, with normal levels of platelets. The anemia was non-regenerative, and a left shift was present. The Coombs' test confirmed that there were autoantibodies present on the erythrocyte membranes. The ultrasound showed that splenic consumption of opsonized erythrocytes was the likely cause of erythrocyte depletion. Testing for tick-borne diseases (as a potential cause of the IMHA) was performed, and the results were negative.

DIAGNOSIS

Brody was diagnosed with immune-mediated hemolytic anemia based on the severe anemia and the positive Coombs' test. The elevated bilirubin concentration supported the conclusion that erythrocyte destruction had occurred.

TREATMENT

Brody received a blood transfusion, which increased his hematocrit to 23%, and he was started on intravenous dexamethasone 0.3 mg/kg every 24 hours, and famotidine 0.5 mg/kg every 12 hours. He also received ondansetron (Zofran) 0.5 mg/kg every 12 hours subcutaneously to prevent vomiting. His clotting time was monitored, and heparin was administered to prevent intravascular coagulation as required. Cyclosporin (50 mg orally every 12 hours) was added to the treatment regimen, and dexamethasone was replaced with prednisone (15 mg orally every 12 hours) once vomiting had been controlled. Brody's reticulocyte and erythrocyte counts continued to increase, and he was sent home with cyclosporin, prednisone, and Pepcid. His owner was advised to weigh the importance of booster vaccinations for the common canine infectious diseases against the possibility of further stimulating an autoimmune response.

IMMUNE-MEDIATED HEMOLYTIC ANEMIA

Immune-mediated hemolytic anemia (IMHA) is one of the most common autoimmune diseases seen in dogs. The disease often presents with an acute onset of anemia, either from erythrocyte lysis or from erythrophagocytosis. Icterus, pallor, and a low hematocrit are characteristic features. Often the patient also presents with vomiting, anorexia, weakness, and lethargy. IMHA is generally a mid-life disease, with onset most commonly occurring at around 4 to 5 years of age. There is also a genetic predisposition. Although IMHA is seen in many breeds, including crossbred dogs, the disease has a higher prevalence among Cocker Spaniels, Old English Sheepdogs, Miniature Dachshunds, and Samoyeds.

IMHA can be classified into different clinical presentations by performing a Coombs' test at either 4°C or 37°C and by using IgG- or IgM-specific antiglobulin. For example, agglutination at 4°C is associated with cyanosis of extremities in cold weather, as erythrocyte agglutination in peripheral vessels impedes oxygenation of the tissue. The ears and tail tip may become necrotic. By contrast, an IgM antibody measured at 37°C is likely to cause intravascular hemolysis. When IgG antibodies, usually IgG1 or IgG4, bind to antigenic determinants on erythrocytes and are not able to bridge between two cells, they may not cause agglutination, and may not fix complement. These antibodies cause anemia by opsonizing the erythrocytes, which are subsequently eliminated by the spleen.

Brody's anemia was classified as primary idiopathic IMHA because it did not appear to be associated with an inciting cause. However, IMHA can be secondary to infection (such as ehrlichiosis or other tick-borne diseases), neoplasia, or chronic inflammation. Drugs can act as haptens, and after binding to erythrocytes they can initiate IMHA. Some examples of drugs that can trigger IMHA include trimethoprim/sulfamethoxazole, Dilantin (phenytoin), and penicillin.

COMPARATIVE MEDICINE CONSIDERATIONS

IMHA occurs in a variety of species, including humans, horses, cattle, cats, and parrots. The condition is not common in human patients; it affects 1 to 3 people per 100,000. The term "immune-mediated hemolytic anemia" is also generally preferred to "autoimmune-mediated hemolytic anemia," because many of these cases can be attributed to immune responses to drugs, such as penicillin or trimethoprim/sulfamethoxazole. Such immune responses are transient. In human patients, IMHA has been reported to occur secondary to certain autoimmune diseases, such as systemic lupus erythematosus, and with some malignancies. It can also be initiated by certain infectious diseases; for example, about 30% of people infected with *Mycoplasma pneumoniae* develop a transient anti-erythrocyte antibody response and associated anemia. However, it is estimated that around 50% of the cases in humans are idiopathic—that is, without a known cause. Many of these are probably related to a breakdown in self tolerance.

Many of the same drugs that are linked to IMHA in humans have been linked to the disease in horses. Several bacterial infections in horses have also been associated with IMHA. *Rhodococcus equi* has been associated with the condition in foals, and *Clostridium perfringens* infection of horses has been reported to be accompanied by Coombs' positive anemia. IMHA is less common in cats than in dogs. However, primary idiopathic IMHA does occur in the cat. The Coombs' test is reported to be less accurate in defining the syndrome in cats than in other species, with many false positives. The anemia is nonregenerative with leukocytosis, and occurs in cats of all ages, with a mean age of onset of 2 years, according to a published study of 19 cases. One study of anemic cattle showed that 30% of 42 anemic cattle had a positive Coombs' test result. Slightly more than 50% of these cattle were either being treated with a drug or had a concurrent infection. Characteristics of the anemia included basophilic stippling, spherocytosis, hyperfibrinogenemia, left shift, and hyperbilirubinemia. Anaplasmosis has also been associated with Coombs'-positive anemia in cattle.

In summary, the development of antibodies that react against erythrocytes occurs in a variety of species, with similar type II immune pathogenesis. IMHA can be secondary to other disease, or to drug administration, or it can be idiopathic.

Questions

1. Some practitioners perform splenectomy on dogs that have IMHA with splenomegaly. What is the basis for this treatment and how might it help to resolve the disease? Which antibody isotype is most frequently associated with extravascular hemolysis?

2. What is the difference between a primary idiopathic IMHA and a secondary IMHA? How would the treatment of these differ?

3. Which antibody isotype is most commonly associated with intravascular hemolysis? Why is this so?

4. Why do you think Brody's veterinarian advised the dog's owner to limit any unnecessary vaccination?

Further Reading

Goggs R, Dennis SG, Di Bella A et al. (2015) Predicting outcome in dogs with primary immune-mediated hemolytic anemia: results of a multicenter case registry. *J Vet Intern Med* 29:1603–1610.

Johnston MS, Son TT & Rosenthal KL (2007) Immune-mediated hemolytic anemia in an eclectus parrot. *J Am Vet Med Assoc* 230:1028–1031.

Kohn B, Weingart C, Eckmann V et al. (2006) Primary immune-mediated hemolytic anemia in 19 cats: diagnosis, therapy, and outcome (1998–2004). *J Vet Intern Med* 20:159–166.

McConnico RS, Roberts MC & Tompkins M (1992) Penicillin-induced immune-mediated hemolytic anemia in a horse. *J Am Vet Med Assoc* 201:1402–1403.

Nassiri SM, Darvishi S & Khazraiinia P (2011) Bovine immune-mediated hemolytic anemia: 13 cases (November 2008–August 2009). *Vet Clin Pathol* 40:459–466.

Swann JW, Szladovits B & Glanemann B (2016) Demographic characteristics, survival and prognostic factors for mortality in cats with primary immune-mediated hemolytic anemia. *J Vet Intern Med* 30:147–156.

Warman SM, Murray JK, Ridyard A et al. (2008) Pattern of Coombs' test reactivity has diagnostic significance in dogs with immune-mediated haemolytic anaemia. *J Small Anim Pract* 49:525–530.

CASE 20
EVANS SYNDROME

There are several potential causes of development of autoimmunity and production of antibodies against self antigens. Sometimes consumption of certain drugs, including antibiotics, can induce production of antibodies against self tissues if a drug metabolite (acting as a hapten) binds to a cell and creates a new antigenic epitope. In many cases the inciting cause of the autoimmune response is not obvious. In immune-mediated anemia, the production of antibodies against erythrocyte antigens causes red blood cell destruction by complement-mediated lysis and/or removal by phagocytic cells (Figure 20.1). The type II hypersensitivity mechanism involved utilizes the classical pathway of complement fixation for lysis of the erythrocytes. Production of C3b, which can act as an opsonin, further facilitates the erythrocyte loss by phagocytic cell

TOPICS BEARING ON THIS CASE:

Humoral autoimmunity

Type II hypersensitivity mechanism for anemia

Coombs' test and anti-megakaryocyte antibody detection

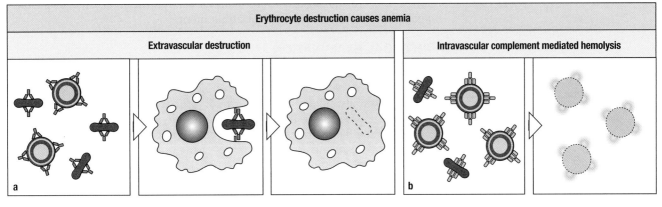

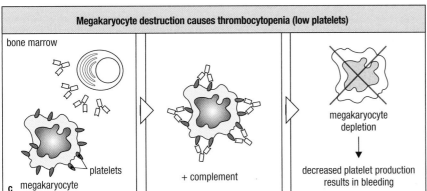

Figure 20.1 Evans syndrome occurs when antibodies are made against both erythrocytes and platelets/megakaryocytes. Both cell types are destroyed by a type II hypersensitivity reaction. Antibody-coated erythrocytes are removed from the circulation by fixed macrophages in the spleen (a) or lysed by complement (b), resulting in anemia. Antibodies against megakaryocytes, the source of thrombocytes, bind to those cells in the bone marrow and cause complement-mediated destruction. Insufficient numbers of thrombocytes result in prevention of blood clotting and cause persistent bleeding (c).

engulfment and destruction. Erythrocytes can also be lost via opsonization with antibody and removal of the antibody-coated cells through Fc-receptor binding by fixed phagocytic cells in the spleen. Whether there is lysis of erythrocytes through fixation of complement or removal of opsonized erythrocytes by phagocytosis, the result is anemia. Autoimmune destruction of thrombocytes (platelets) and/or megakaryocytes, the parental cells in the bone marrow, results in low platelet counts and blood loss from coagulation deficiency. The binding of anti-megakaryocyte antibodies to megakaryocytes in the bone marrow initiates megakaryocyte destruction, with a resultant decrease in production of platelets. Autoimmune destruction of erythrocytes and autoimmune destruction of thrombocytes occur as independent diseases, but can occur together in a condition called Evans syndrome.

THE CASE OF FANCY: A HORSE WITH UNSTOPPABLE BLEEDING FROM THE NOSE

SIGNALMENT/CASE HISTORY

Fancy, a 6-year-old American Quarter Horse mare, lives in a pasture where there are blackberry bushes. One morning her owner found her profusely bleeding from both nostrils. Thinking that she had simply scratched her nose on a bush, he applied pressure and cleaned up her nose. However, after several hours the mare's nose was continuing to bleed, and the owner called the veterinarian. Prior to this episode the mare appeared to be acting normally, although approximately 1 month previously she had been treated for a dry cough with the antimicrobial trimethoprim/sulfamethoxazole. Fancy's vaccines were up to date and she was on a regular deworming schedule. The other four horses in her pasture were healthy. Fancy's diet consisted of pasture, alfalfa hay, and a vitamin mineral supplement.

PHYSICAL EXAMINATION

On physical examination, Fancy's temperature was 101.2°F (normal range, 100–102°F), her pulse was 64 beats/minute (normal range, 30–40 beats/minute), and her respiratory rate was normal at 20 breaths/minute. Her capillary refill time was normal at 1–2 seconds. All of her mucous membranes were pale, and petechial hemorrhages were present. Both hind legs showed evidence of edema, and digital pulses were increased. Fancy's submandibular lymph nodes were enlarged, and blood was seen coming from both nares (Figure 20.2). The packed cell volume (PCV) was determined to be 12 (normal range, 32–53).

DIFFERENTIAL DIAGNOSIS

Pale mucous membranes and low PCV are indicative of anemia. There are a variety of causes of anemia, including decreased erythrocyte production due to bone-marrow suppression, erythrocyte destruction (immune mediated or toxic), and erythrocyte loss from bleeding. Nasal tumor, nasal trauma, and pulmonary hemorrhage are three of the most frequent causes of epistaxis. The age of this horse makes a tumor less likely, and pulmonary hemorrhage is most often seen in race horses after exertion. The presence of petechiae indicates that it is likely there is a problem with Fancy's blood-clotting mechanism. This could involve a variety of clotting factors as well as low platelet numbers and/or impaired function.

DIAGNOSTIC TESTS AND RESULTS

A complete blood count (CBC) and a blood chemistry and clotting panel were performed. The most notable results from the CBC were anemia (PCV, 12; normal range, 32–53), thrombocytopenia (platelet count, 0; normal range,

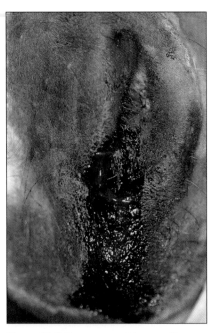

Figure 20.2 Fancy, several hours after her owner first noticed the bleeding. Fresh blood can be seen coming from the left nare. This is evidence of epistaxis. (Courtesy of Derek Knottenbelt.)

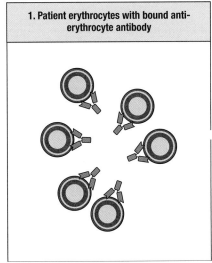

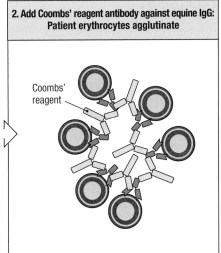

1. Patient erythrocytes with bound anti-erythrocyte antibody

2. Add Coombs' reagent antibody against equine IgG: Patient erythrocytes agglutinate

Coombs' reagent

Figure 20.3 The agglutination reaction seen in a positive Coombs' test. Patient erythrocytes with antibody bound to the surface and cross-linked or agglutinated by addition of "Coombs reagent", an antibody to equine IgG. The test detects non-agglutinating, non-complement-fixing antibodies on erythrocytes. The red/blue antibodies show anti-erythrocyte IgG made by the horse against self cells (1); the green antibodies are a reagent antibody that binds to equine IgG (2). Higher titers reflect larger amounts of anti-erythrocyte antibody. Erythrocytes that do not have equine antibody bound to the surface will not be agglutinated by the anti-equine antibodies, as there is nothing for them to bind to, and the test will be negative.

100,000–300,000/μL), and hypoproteinemia (4.2 g/dL; normal range, 5.8–8.7 g/dL). The clotting panel showed the prothrombin time to be low at 12.2 seconds (normal range, 14–16 seconds). The PIVKA assay, which determines whether there has been a reduction in vitamin K levels by detecting production of a protein induced by the vitamin's absence, was normal. Some poisons cause defective blood clotting by reducing vitamin K levels. The antithrombin III assay was also normal. Reduced levels of antithrombin III can result in enhanced blood coagulation. The direct Coombs' test (Figure 20.3), which evaluates the presence or absence of anti-erythrocyte antibodies bound to the surface of the patient's erythrocytes, was positive with a titer of 32.

Microscopic examination of the bone-marrow aspirate showed an increase in megakaryocytes. Myeloid and erythroid cell lines showed normal differentiation, with frequent mitotic figures. Rare plasma cells were also seen. Anti-megakaryocyte antibody testing was positive, indicating the presence of autoantibodies bound to megakaryocytes. In this test a fluorescence marker bound to an antibody against equine antibody is used to detect whether there are antibodies bound to the bone-marrow megakaryocytes. The presence of these antibodies indicates that a type II hypersensitivity reaction is the likely cause of platelet loss. The megakaryocytes showed 2+ to 3+ fluorescence. Positive fluorescence was also observed on small erythrocyte precursors late in maturation. This indicated that antibodies against erythrocyte epitopes were also being produced by the patient.

DIAGNOSIS

The presence of a mucous membrane pallor and petechiae was indicative of both anemia and thrombocytopenia. The low hematocrit, positive Coombs' test, low platelet count, and positive anti-megakaryocyte antibody test confirmed that Fancy had both immune-mediated anemia and immune-mediated thrombocytopenia, which when they occur together are called Evans syndrome.

TREATMENT

The initial treatment in this case involved blood transfusion because Fancy's hematocrit was dangerously low. She was also treated with dexamethasone (a corticosteroid) to suppress the inappropriate immune response. Initially auto-immune anemia and thrombocytopenia are treated with a high dose of corticosteroid, and the hematocrit and platelet counts are checked weekly. Once these parameters have returned to normal the dose can usually be gradually decreased, and sometimes stopped after a period of time. Fancy responded well to treatment, and ultimately was able to stop immunosuppressive therapy.

Often the prognosis for Evans syndrome is grave. However, if the autoimmune response was initiated by a drug, once that drug has been removed from the body there is a greater likelihood that immunosuppression can be stopped.

EVANS SYNDROME

Evans syndrome occurs when both red blood cells and platelets are attacked by autoantibodies. The occurrence of autoimmune hemolytic anemia (immune-mediated anemia) in the horse has been associated with treatment with trimethoprim/sulfamethoxazole in some reported cases, including this one. Although there is no proof of cause, it can be speculated that Fancy's immune system may have been disturbed by the treatment. There is a higher incidence of autoimmune hemolytic anemia (AIHA) and immune-mediated thrombocytopenia (ITP) in horses infected with equine infectious anemia virus. However, the occurrence of both AIHA *and* ITP, as in Evans syndrome, is rare and indicates a profound loss of self tolerance. In one study of 35 horses with thrombocytopenia (regardless of mechanism), only one of the horses also had a Coombs-positive anemia (Evans syndrome), which corresponds to an incidence of 2.85%.

COMPARATIVE MEDICINE CONSIDERATIONS

Evans syndrome is a rare disease that was discovered in 1951 by Robert Evans and his colleagues. They described a group of human patients who had both immune mediated hemolytic anemia (Coombs' positive) and idiopathic thrombocytopenia. Some of the patients also had antibodies to white blood cells. It is estimated that 10–23% of human patients who have autoimmune hemolytic anemia will also have autoimmune thrombocytopenia. As in horses, presenting signs include petechiae or ecchymotic hemorrhages and pallor, indicative of anemia. When the anemia is hemolytic, jaundice is usually present as well. The disease is thought to represent a state of profound immune dysregulation, and is sometimes seen in patients with lymphoproliferative disease or systemic lupus erythematosus. Therapy involves dexamethasone, other more potent immunosuppressive drugs, intravenous immunoglobulin, and, in refractory cases, splenectomy.

Evans syndrome has also been described in dogs. Cocker Spaniels, Bichon Frises, and Pugs in particular have shown susceptibility to this condition.

Questions

1. What was the cause of the low hematocrit (PCV) in this case?

2. What was the cause of the petechiae and the epistaxis in this case?

3. In this case, autoantibodies against erythrocytes were detected using a specific assay, and those against megakaryocytes were detected with another type of test. Describe each of these tests and explain how they differ from each other.

4. If the antibodies detected by the Coombs' test were bound to erythrocytes, why did they fail to cause erythrocyte lysis by complement fixation?

Further Reading

Evans RS, Takahashi K, Duane RT et al. (1951) Primary thrombocytopenic purpura and acquired hemolytic anemia; evidence for a common etiology. *Arch Intern Med* 87:48–65.

McGovern KF, Lascola KM, Davis E et al. (2011) T-cell lymphoma with immune-mediated anemia and thrombocytopenia in a horse. *J Vet Intern Med* 25:1181–1185.

Nunez R, Gomes-Keller MA, Schwarzwald C & Feige K (2001) Assessment of Equine Autoimmune Thrombocytopenia (EAT) by flow cytometry. *BMC Blood Disord* 1:1.

Pegels JG, Helmerhorst FM, van Leeuwen EF et al. (1982) The Evans syndrome: characterization of the responsible autoantibodies. *Br J Haematol* 51:445–450.

Sellon DC, Levine J, Millikin E et al. (1996) Thrombocytopenia in horses: 35 cases (1989–1994). *J Vet Intern Med* 10:127–132.

Thomas HL & Livesey MA (1998) Immune-mediated hemolytic anemia associated with trimethoprim–sulphamethoxazole administration in a horse. *Can Vet J* 39:171–173.

CASE 21
PEMPHIGUS
FOLIACEUS

Pemphigus is a type of autoimmune skin disease in which there is a type II hypersensitivity response against self antigens in the skin, specifically proteins that form part of the intercellular junctions that keep epidermal cells together. The term "pemphigus" refers to blister formation, and all forms of pemphigus have vesicle formation as a characteristic. The autoantibodies involved target members of the cadherin family of cell adhesion molecules. Binding of autoantibodies to these molecules is followed by complement fixation and destruction of epithelial integrity. Autoantibodies target extracellular portions of the desmosomal adhesion proteins between keratinocytes, causing the latter to lose cohesion with each other and resulting in acantholysis. There are several distinct autoimmune skin diseases that involve development of autoantibodies against epidermal adhesion proteins. Clinical severity is a reflection of the depth of the acantholysis. The most severe form is pemphigus vulgaris, in which deep layers of the epidermis are involved. Pemphigus foliaceus (PF) is the most common antibody-mediated autoimmune skin disease of dogs and other domestic animal species.

THE CASE OF ROBBIE: A DOG WHO DEVELOPED BLISTERS AND EROSIONS ON HIS SKIN

SIGNALMENT/CASE HISTORY

Robbie is a 2-year-old castrated male Border Collie. He was presented to his veterinarian for evaluation of crusting skin lesions on his face and foot pads. The lesions had first started to develop 4 months earlier, and began on his dorsal muzzle, with crusting lesions and depigmentation of the nasal planum. The lesions partially and transiently improved when prednisone was administered, but over time they have progressed, and now most of the dorsal muzzle, periocular regions, and foot pads are affected. The lesions are uncomfortable and the dog will rub or lick affected areas. There are no other medical concerns, no previous medical problems, and no prior history of drug administration. Robbie's vaccinations are up to date and he was heartworm negative on a test 6 months earlier, but is not currently receiving any preventative medications.

PHYSICAL EXAMINATION

Robbie's physical examination findings were unremarkable, with vital signs within normal limits. The only abnormalities found involved the skin. There was bilaterally symmetric peri-ocular alopecia, with crusting and signs of erosive lesions from self-trauma removing crusts. The dorsal muzzle was alopecic

TOPICS BEARING ON THIS CASE:

Autoimmunity

Type II hypersensitivity

Immune-mediated skin disease

Figure 21.1 Robbie showing alopecia and crusting on the dorsum of the nose. (Courtesy of Verena Afolter and Catherine Outerbridge.)

Figure 21.2 Reddish discoloration of the hair around the footpads indicates that Robbie has been licking his feet. (Courtesy of Verena Afolter and Catherine Outerbridge.)

with multifocal coalescing crusts and erosive lesions. The nasal planum was depigmented and the adherent crusts extended onto the dorsal aspect of the nasal planum (Figure 21.1). All four feet had crusting lesions involving all of the footpads. There was also reddish discoloration of the hairs on the palmar and plantar aspects of the feet, probably due to chronic licking (Figure 21.2).

DIFFERENTIAL DIAGNOSIS

Differential diagnoses included pemphigus foliaceus, dermatophytosis (in particular *Trichophyton mentagrophytes* infection), demodicosis with secondary infection and self-trauma, superficial necrolytic dermatitis, and zinc-responsive dermatosis.

DIAGNOSTIC TESTS AND RESULTS

Initial diagnostic tests included obtaining samples from the skin to examine for evidence of secondary infections or acantholytic cells. *Demodex* and *Trichophyton* infections were ruled out by negative skin scrapings. Cytologic review revealed coccoid bacteria, degenerated and non-degenerated neutrophils, and acantholytic cells. Although the presence of acantholytic cells may be consistent with pemphigus foliaceus, histologic examination of skin biopsies taken from areas with intact pustules or adherent crusts is needed to confirm the diagnosis. Multiple skin biopsies were taken from a thick area of adherent crusting along the paw pad margin, an area of crusting above the eye, and an area of adherent crusting on the dorsal muzzle. The histologic examination revealed features consistent with the diagnosis of pemphigus foliaceus.

DIAGNOSIS

Robbie was diagnosed with pemphigus foliaceus based on the gross lesions and the histopathology of skin biopsy showing typical features of the disease, namely intra-epidermal pustular dermatitis, with acantholytic keratinocytes, caused by disruption of the desmosomes.

TREATMENT

Oral prednisolone therapy was initiated at a dose of 2.3 mg/kg/day. Azathioprine was also prescribed as a corticosteroid-sparing agent to achieve immunosuppressive therapy that would allow for more rapid tapering of the prednisolone. Over the next 2 years, treatment regimens changed as attempts to control the lesions of pemphigus foliaceus had to be balanced against minimizing the side effects from the prescribed therapies. Ultimately Robbie's symptoms were managed with alternate-day cyclosporine, low-dose methyl-prednisolone, and topical betamethasone.

PEMPHIGUS FOLIACEUS

The diagnostic features of pemphigus foliaceus in dogs include an intra-epidermal pustular dermatitis with numerous neutrophils and acantholytic keratinocytes, produced by disruption of the desmosomes. The pustules tend to span several follicular openings (Figure 21.3A). Typically the intralesional neutrophils are non-degenerate, and the acantholytic cells are seen as single cells as well as in rafts. Pemphigus foliaceus lesions tend to occur in waves—thus it is common to see layered crusts composed of parakeratosis and hyperkeratosis alternating with serocellular crusting (Figure 21.3B). Crusts often contain acantholytic keratinocytes and degenerated neutrophils (Figure 21.3C). Occasionally, these crusts are overlying new pustules (Figure 21.3D). There is a perivascular to interstitial mixed dermal infiltrate associated with dermal edema, dilated vessels, and congestion. Neutrophils and occasionally eosinophils predominate in the inflammatory infiltrate. With transition into subacute lesions, there are more plasma cells and

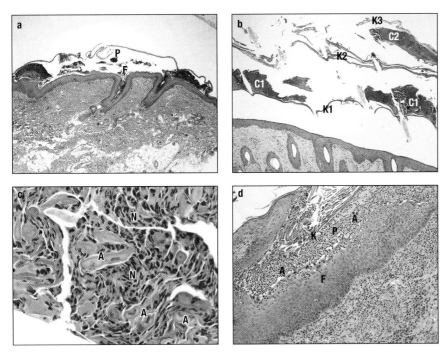

Figure 21.3 (a) A large intact subcorneal pustule (P) spans several hair follicle openings (F). (b) Pemphigus foliaceus often tends to occur in waves of acute disease. This is reflected in the layered crusts: keratin (K1) – crust (C1) – keratin (K2) – crust (C2) – keratin (K3). (c) Within the crusts there are numerous degenerated neutrophils (N) and numerous acantholytic keratinocytes (A). (d) The process of acantholysis and formation of subcorneal pustules can extend into the hair follicles. Between the follicular epithelium (F) and the keratin layer of the follicle (K) is a pustule (P) composed of many neutrophils and scattered acantholytic keratinocytes (A). (Courtesy of Verena Afolter and Catherine Outerbridge.)

lymphocytes present. Facial lesions, especially those close to the nasal planum, may be characterized by an increased number of inflammatory cells between the epidermis and the dermis (lichenoid infiltrate).

IgG, frequently C3, and less commonly IgM deposition may be found in the intercellular spaces of the epidermis after staining by immunohistochemistry or immunofluorescence. However, careful evaluation is warranted, as immunoglobulin deposition can also be seen in other inflammatory lesions. Staining for desmoglein-1 reveals a change of the staining pattern in 62% of dogs with pemphigus foliaceus. Instead of the uniform finely granular delineation found in normal keratinocytes, peripheral and/or dark cytoplasmic staining of keratinocytes is seen, characterized by streaked or perinuclear deposits. Serum of dogs with pemphigus foliaceus contains anti-desmocollin-1 antibodies and to a lesser degree anti-desmoglein-1 antibodies; these bind to the desmosomal areas, as can be detected by indirect immunofluorescence on monkey esophagus cells with the patient's serum. Acantholytic cells can also be seen secondary to the effects of proteolytic enzymes released by degenerating neutrophils. For this reason it is important, particularly in cases where there is an absence of rafts of acantholytic cells, to rule out bacterial or fungal pustular dermatitis by either culture or special stains. The latter include the periodic acid–Schiff (PAS) stain, Gomori methenamine–silver nitrate stain, and Brown and Brenn (B&B) Gram stain.

COMPARATIVE MEDICINE CONSIDERATIONS

In cats and horses, the presence of fresh pustules with rafts of acantholytic cells is less common. It is crucial to have the adherent crusts available in the biopsy to demonstrate the presence of the acantholytic cells within the crusts.

The exact desmosomal protein targeted by autoantibodies in pemphigus foliaceus has not been identified in horses or cats.

In humans, pemphigus foliaceus is predominantly a subcorneal vesicular disease, characterized by an absence of inflammatory cells accompanying the acantholytic cells. The primary protein targeted in humans is desmoglein-1 and not desmocollin-1. In both humans and dogs there appears to be a genetic association with development of the disease.

Questions

1. In what respects are myasthenia gravis (Case 22) and pemphigus foliaceus similar conditions?

2. How do the antibodies against desmoglein-1 cause cell death and blister formation in dogs affected with pemphigus foliaceus?

3. Immunosuppressive therapy is usually sufficient to bring the blistering skin lesions under control in this disease. However, development of secondary infection is common. Why might this be?

Further Reading

Bizikova P, Dean GA, Hashimoto T & Olivry T (2012) Cloning and establishment of canine desmocollin-1 as a major autoantigen in canine pemphigus foliaceus. *Vet Immunol Immunopathol* 149:197–207.

Ishii K, Amagai M, Hall RP et al. (1997) Characterization of autoantibodies in pemphigus using antigen-specific enzyme-linked immunosorbent assays with baculovirus-expressed recombinant desmogleins. *J Immunol* 159:2010–2017.

Olivry T, LaVoy A, Dunston SM et al. (2006) Desmoglein-1 is a minor autoantigen in dogs with pemphigus foliaceus. *Vet Immunol Immunopathol* 110:245–255.

Olivry T, Dunston SM, Walker RH et al. (2009) Investigations on the nature and pathogenicity of circulating antikeratinocyte antibodies in dogs with pemphigus foliaceus. *Vet Dermatol* 20:42–50.

Vandanabeele SI, White SD, Affolter VK et al. (2004) Pemphigus foliaceus in the horse: a retrospective study of 20 cases. *Vet Dermatol* 15:381–388.

The neuromuscular junction (NMJ) is a well-characterized synapse with defined pre- and post-synaptic components that express various ion channels, including the acetylcholine receptor (AChR). Neuromuscular diseases associated principally with autoantibodies to various antigenic targets, including ion channels, can be divided broadly into two groups—those in which the antibodies are markers for an inflammatory process, and those in which the antibodies are directly pathogenic. The antigenic target in immune-mediated myasthenia gravis (MG) is the AChR, and AChR autoantibodies are directly pathogenic. Muscular weakness in myasthenia gravis is caused by loss of AChRs at the neuromuscular junction (NMJ), compounded by morphological damage to the motor endplate. The mechanisms by which autoantibodies lead to loss of AChRs and blockage of nerve impulses to the muscle include direct blockade of receptor function, increased internalization and degradation of AChRs at the NMJ, and complement-mediated damage to the NMJ (Figure 22.1). It is not known what induces the autoimmune response to muscle AChRs in myasthenia gravis.

THE CASE OF DERBY: A DOG WITH PROGRESSIVE WEAKNESS AND REGURGITATION

SIGNALMENT/CASE HISTORY

Derby, a 1-year-old female spayed Great Dane cross, was presented for evaluation of acute ascending weakness progressing from the pelvic limbs to the thoracic limbs over a 24-hour period. She was previously healthy and her vaccinations were up to date. There was no known history of exposure to toxins or ticks, or recent travel.

PHYSICAL EXAMINATION

The patient was alert but not ambulatory, with a body condition score of 4 out of 9,[1] and with symmetrical muscling. Saliva was observed on the thoracic limbs. Clear lung sounds were noted in all fields, and no murmurs were detected. On neurological examination, cranial nerve and spinal reflexes, voluntary movement, and pain perception were found to be intact. Over the next 12 hours, regurgitation of food and water was noted.

TOPICS BEARING ON THIS CASE:

Humoral autoimmunity

Type II hypersensitivity mechanism of immune-mediated damage

Neuromuscular junction

---

[1] According to the 2010 AAHA Nutritional Assessment Guidelines for Dogs and Cats.

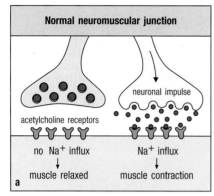

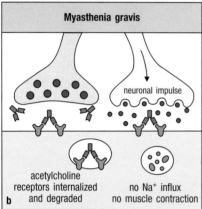

Figure 22.1 (a) The normal function of the neuromuscular junction. Acetylcholine released from neurons binds to receptors on the muscle. This stimulates contraction. (b) Acetylcholine receptors that have been destroyed by binding of autoantibodies specific to the receptors; muscle contraction does not occur. (From Geha R & Notarangelo L [2016] Case Studies in Immunology, 7th ed. Garland Science.)

DIFFERENTIAL DIAGNOSIS

Differential diagnoses for acute ascending neuromuscular weakness include polyradiculoneuritis, tick paralysis, acute idiopathic polyneuropathy, and acute myasthenia gravis. The lack of exposure to ticks made tick paralysis a less likely diagnosis.

DIAGNOSTIC TESTS AND RESULTS

A complete blood count (CBC) was unremarkable. The only abnormality identified on a serum chemistry panel was an elevated creatine kinase (CK) activity of 1200 IU/L (normal range, 50–400 IU/L). Total thyroxine (T4) was within the normal range. Blood gas measurements, lactate, glucose, and electrolytes measured in the ICU were also within the normal range.

An intravenous edrophonium chloride challenge test resulted in an increase in muscle strength, consistent with a diagnosis of immune-mediated myasthenia gravis. The diagnosis of myasthenia gravis was confirmed by detection of autoantibodies against muscle AChRs by immunoprecipitation radioimmunoassay with a titer of 1.30 nmol/L (canine normal range, < 0.6 nmol/L). Dogs with myasthenia gravis frequently have decreased muscle contraction in the esophagus, which can cause regurgitation of food, and ultimately megaesophagus with subsequent aspiration pneumonia. Although thoracic radiographs did not demonstrate obvious esophageal dilatation, a mild to moderate amount of gas and fluid was present in the thoracic esophagus, consistent with esophageal dysmotility.

DIAGNOSIS

Derby was diagnosed with immune-mediated myasthenia gravis based on detection of autoantibodies against muscle AChRs by immunoprecipitation radioimmunoassay. This assay remains the gold standard for the diagnosis of acquired myasthenia gravis. It involves precipitation of serum IgG and IgM antibodies that bind to solubilized AChR complexed with the high-affinity peptide agonist, I^{125}-labeled α-bungarotoxin. The precipitate's γ-ray emission is a measure of the amount of AChR bound to immunoglobulin. The assay is specific and sensitive, and positivity demonstrates an autoimmune response against muscle AChRs. Within an individual, AChR antibody levels correlate well with disease severity, but antibody levels between patients are highly variable and do not correlate well with severity. This suggests the possibility that there are individual differences in the specificity of the antibodies, or the existence of antibodies to other NMJ antigens. Binding of AChR antibody to AChRs can often be visualized using immunoenzyme staining techniques.

TREATMENT

Altered feeding procedures, including elevation of food and water, were employed due to the potential for aspiration pneumonia (Figure 22.2). Anticholinesterase drugs are the cornerstone of therapy for acquired myasthenia gravis; they prolong the action of acetylcholine at the NMJ and enhance neuromuscular transmission. Pyridostigmine bromide was prescribed, and an improvement in muscle strength was noted. Since an optimal response to therapy was obtained with anticholinesterase drugs alone, resulting in the return of normal limb muscle strength, supportive care and anticholinesterase drugs were all that was required for therapy. If an optimal response had not been obtained, low-dose corticosteroids or other immunosuppressive drugs might have been added.

ACQUIRED MYASTHENIA GRAVIS

In dogs, acquired myasthenia gravis is differentiated from familial myasthenia gravis, which is caused by autosomal recessive inheritance in several dog breeds, including Jack Russell Terriers, Smooth Fox Terriers, Miniature

Dachshunds, and English Springer Spaniels. In the congenital disease the defect is manifested at an early age. In contrast, acquired myasthenia gravis usually occurs later in life, and there are multiple factors involved in initiation of the disease. As with other autoimmune diseases, there is a genetic background with which other factors—such as environment, hormones, and infections—interact in the elicitation of the loss of self tolerance. Several breeds have an increased incidence of the acquired form of myasthenia gravis, including the Newfoundland and the Great Dane.

COMPARATIVE MEDICINE CONSIDERATIONS

As in canines, progressive weakness is a common presentation in human patients with myasthenia gravis. Unlike canines, human patients may present with weakness in the ocular muscles, causing ptosis of the eyelids and slow ocular movement. Canine patients with advanced myasthenia gravis often develop megaesophagus, which results in regurgitation and aspiration pneumonia. Human patients can develop difficulty chewing, and subsequently may also develop aspiration pneumonia. In both species, there have been cases reported associated with thymoma (tumor of the thymus). Treatment with pyridostigmine is similar in both species, with the addition of immunosuppressive drugs when needed. Plasmapheresis (also called therapeutic plasma exchange) has been used to treat refractory human cases of myasthenia gravis, and has recently been used for some canine patients, although it is currently an extremely expensive mode of therapy.

Myasthenia gravis is also seen in cats, in which generalized weakness is the most common presentation. Megaesophagus is not usually present, but the presence of a cranial mediastinal mass is common. The incidence is higher in Abyssinian and Somali cats than in the rest of the cat population. There has also been a reported case of myasthenia gravis in a 10-year-old Siberian tiger. The tiger had a T-cell-rich thymoma, and on necropsy testing of muscle tissue showed the presence of bound anti-AChR antibodies.

Figure 22.2 Derby showing signs of extreme muscle weakness. The apparatus in which she has been placed is designed to allow the affected dog to be vertical while eating, helping to counteract the regurgitation that can occur with decreased muscle contraction in the esophagus, and prevent megaesophagus with subsequent aspiration pneumonia. (Courtesy of the Karen Verneau Neurology/Neurosurgery Service at the UC Davis Veterinary Medical Teaching Hospital.)

Questions

1. The muscular weakness in myasthenia gravis that results from destruction of AChRs by antibodies can be alleviated by administration of pyridostigmine bromide. How does this drug improve motor function?

2. Derby did not require treatment with corticosteroids, but some patients do not respond to pyridostigmine bromide alone. What is the mechanism by which the corticosteroid stops the progression of disease?

3. In human patients, plasmapheresis is performed in patients with severe disease. What is the purpose of this process?

Further Reading

Allan K, Masters N, Rivers S et al. (2014) T-lymphocyte-rich thymoma and myasthenia gravis in a Siberian tiger (*Panthera tigris altaica*). *J Comp Pathol* 150:345–349.

Dewey CW, Bailey CS, Shelton GD et al. (1997) Clinical forms of acquired myasthenia gravis in dogs: 25 cases (1988–1995). *J Vet Intern Med* 11:50–57.

Hague DW, Humphries HD, Mitchell MA & Shelton GD (2015) Risk factors and outcomes in cats with acquired myasthenia gravis (2001–2012). *J Vet Intern Med* 29:1307–1312.

Lipsitz D, Berry JL & Shelton GD (1999) Inherited predisposition to myasthenia gravis in Newfoundlands. *J Am Vet Med Assoc* 215:956–958.

Rusbridge C, White RN, Elwood CM & Wheeler SJ (1996) Treatment of acquired myasthenia gravis associated with thymoma in two dogs. *J Small Anim Pract* 37:376–380.

Shelton GD (2010) Routine and specialized laboratory testing for the diagnosis of neuromuscular diseases in dogs and cats. *Vet Clin Pathol* 39:278–295.

CASE 23
SYSTEMIC LUPUS ERYTHEMATOSUS

Immune complexes form between soluble antigen and circulating antibody whenever both are present within the blood. Removal of these complexes by fixed macrophages in the spleen and liver is the body's effective way of removing antigen, and when the antigen is from an infectious disease agent this is beneficial to the patient. However, sometimes the size of the immune complexes causes them to become entrapped in small blood vessels, such as those in the kidney glomeruli, skin, and joint synovium. Complement is readily fixed when either IgM or IgG immune complexes are made. Fixed complement in turn releases chemotactic factors (C3a and C5a), which draw neutrophils to the area. Joints that are affected by immune-complex deposition therefore have large numbers of neutrophils in the synovial fluid. Glomeruli that are affected by immune-complex deposition eventually leak protein into the urine, because the release of enzymes (such as proteases and collagenases) from the neutrophils into the area of the complexes damages the capillary endothelium. The resulting inflammation causes a vasculitis, which is a characteristic of type III hypersensitivity (Figure 23.1). This process occurs in the syndromes known as serum sickness and the Arthus reaction, and also in the systemic autoimmune disease known as systemic lupus erythematosus (SLE).

THE CASE OF BOVER: A DOG FOR WHOM SIMPLY GETTING UP AND DOWN BECAME TOO PAINFUL

SIGNALMENT/CASE HISTORY

Bover is a 3-year-old male English Pointer who presented to his veterinarian with a history of increasing weakness and difficulty rising. His owner reported that Bover had become very lethargic during the last 6 weeks, and had recently developed a very red and sore area across the top of his nose. He was still eating and drinking normally, and had no additional complaints. Bover lives in a suburban home and has access to a grassy yard, but does not venture into woodland and has never been exposed to ticks (to the owner's knowledge). He often sits on the back porch in the sun. His core vaccines are up to date, and he is on a regular regime of Heartgard® and Frontline®. Prior to his bout of lethargy he would frequently play with other dogs at the local dog park. His owner reported that Bover was never left unsupervised, and to his knowledge had no history of trauma.

TOPICS BEARING ON THIS CASE:

Immune-complex disease

Type III hypersensitivity

Autoimmunity

Antinuclear antibodies

	Type III
Immune reactant	IgG
Antigen	soluble antigen
Effector mechanism	complement, phagocytes
	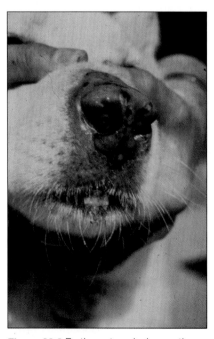 immune complex + complement
Example of hypersensitivity reaction	Serum sickness, Arthus reaction

Figure 23.1 Type III hypersensitivity is caused by a combination of soluble antigen and IgG antibodies. Subsequent complement fixation in the walls of small blood vessels creates chemotactic molecules, which elicit neutrophil migration. Degranulation of neutrophils in and around the tissue causes vasculitis. (From Murphy K [2011] Janeway's Immunobiology, 8th ed. Garland Science.)

Figure 23.2 Erythematous lesion on the dorsum of Bover's nasal planum, and erythema of the surrounding skin.

PHYSICAL EXAMINATION

On physical examination, Bover had a temperature of 102.5°F (39.2°C) (normal range, 100.5–102°F). His heart and lungs were normal on auscultation. His body condition was average,[1] but there was significant muscle atrophy present on both rear and forelimbs. The carpi and tarsi showed swelling bilaterally, but were not painful. The peripheral lymph nodes were enlarged. Bover also had an erythematous lesion on the dorsum of his nose (Figure 23.2).

DIFFERENTIAL DIAGNOSIS

The primary presenting clinical sign was multi-limb lameness of significant duration without any history of trauma. The presence of enlarged lymph nodes increased the likelihood that the lameness had a systemic cause. Possible diagnoses include Lyme disease, rheumatoid arthritis, and immune-complex-mediated polyarthritis (alone or as part of SLE). The lesion on the nose could be due to trauma, such as might occur if Bover was repeatedly poking his nose through a fence opening. The location and history of sun exposure could indicate discoid lupus, or the lesion could be a manifestation of SLE.

DIAGNOSTIC TESTS AND RESULTS

A complete blood count (CBC), chemistry panel, and urinalysis were performed. The CBC showed a low normal red blood cell count and hematocrit ($5.7 \times 10^6/\mu L$ and 35%, respectively), an increased plasma protein concentration (8.0 g/dL), and a high white blood cell count ($18 \times 10^3/\mu L$). On the chemistry panel the albumin/globulin (A/G) ratio was decreased at 0.72. Joint taps were performed on multiple joints on both the right and left legs (Table 23.1). These results indicated the presence of a high number of neutrophils within the joint fluid. No organisms were seen on a direct slide of the aspirate, but neutrophils could be visualized. Culture of the synovial fluid did not grow any bacteria. In summary, the joint taps showed the presence of inflammatory cells in the joints without infection, and were therefore consistent with immune-mediated inflammation. Urinalysis results (obtained by cystocentesis) were normal, with no protein detected. An antinuclear antibody (ANA) test performed on serum was positive at a dilution of 1:640, and the pattern was homogeneous (Figure 23.3). This ANA test pattern is associated with canine SLE. A biopsy of the lesion on the dorsum of the nose was stained with fluorescein-conjugated anti-C3 and anti-IgG; the results showed the typical "lupus band" at the dermal–epidermal junction (Figure 23.4). This indicated the deposition of antibody–antigen complexes at the dermal–epidermal junction, and if C3 was present, the fixation of complement by those complexes.

Table 23.1 Results of initial joint taps (normal values are 0; there should not be any inflammatory cells in the joint)

Site of joint fluid collection	Total white blood cells/mm³	Neutrophils (% of cells)	Small mononuclear (% of cells)	Large mononuclear (% of cells)
Left carpus	5200	40	7	53
Left hock	4400	86	5	9
Left stifle	12,000	35	13	52
Right carpus	2600	42	6	52
Right hock	5200	87	2	11
Right stifle	25,000	77	2	21

---

1 According to the Purina® body condition rating system.

DIAGNOSIS

The diagnosis of systemic lupus erythematosus requires that there is a positive ANA test (titer > 100) or a positive lupus erythematosus (LE) test, and the involvement of at least two body systems. Bover had an ANA titer of 640 with involvement of joints and skin, thus fulfilling the diagnosis criteria.

TREATMENT

Bover was started on a regimen of prednisone and the immunosuppressive drug Imuran® (azathioprine). His condition improved, as demonstrated by a reduction in his lameness, decreased numbers of neutrophils in the joint taps, and a decrease in ANA titer (by 4 months post diagnosis the ANA titer had declined to 40). The dose of prednisone was gradually decreased to a maintenance dose, and the Imuran® was discontinued while the dog remained in remission. The lesion on the nasal planum improved, and the owner was advised to limit sun exposure and to apply sunscreen when the dog was outside.

SYSTEMIC LUPUS ERYTHEMATOSUS

SLE is a multisystem autoimmune disease characterized by the production of antibodies to components of cell nuclei, primarily double-stranded DNA and histones. It is currently thought that malfunction of apoptosis of cells occurs in patients with SLE. The immune complexes that initiate the vasculitis result from binding of nuclear antigens with the antinuclear antibodies. Patients may present with polyarthritis, proteinuria (immune-complex glomerulonephritis), erosive skin lesions (particularly on the face, and often exacerbated by exposure to ultraviolet light), Coombs'-positive anemia, and immune-mediated thrombocytopenia. Using the criteria developed for SLE in human patients, which have subsequently been adopted for canine patients, at least two of these conditions must be present in addition to a positive ANA or LE cell preparation. Another common characteristic of SLE is the presence of a polyclonal gammopathy due to the activation of multiple B cells, causing an increase in production of immunoglobulin and a heterogeneous accumulation of the latter, in contrast to the homogeneous immunoglobulin production that is characteristic of a monoclonal gammopathy (Figure 23.5). Although a low albumin/globulin ratio in the serum can indicate a polyclonal gammopathy, it may also be caused by low albumin levels resulting from kidney or liver disease.

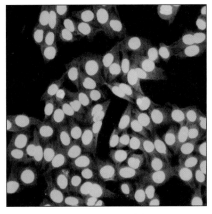

Figure 23.3 Antinuclear antibody test showing homogeneous immunofluorescence within HEp-2 cell nuclei. This test is performed by incubating dilutions of the patient's serum on slides containing fixed HEp-2 cells. If antinuclear antibodies are present in the serum they bind to their target antigens (such as DNA and histones) and cause the nuclei of the HEp-2 cells to fluoresce when a fluorescein-conjugated antibody against dog IgG is added. A titer is determined by evaluation of the different spots that were reacted with the patient's diluted serum. The last spot that shows a positive reaction is considered to be the titer.

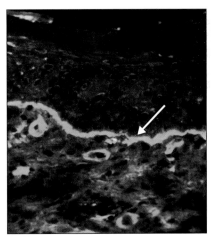

Figure 23.4 Immunofluorescence of skin biopsy from the nasal planum stained with FITC-conjugated anti-canine IgG showing antibody deposition forming a typical "lupus band" at the dermal–epidermal junction (indicated by arrow).

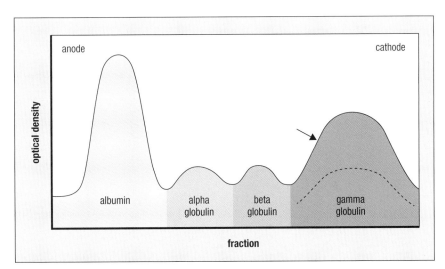

Figure 23.5 Densitometry tracing from serum electrophoresis showing polyclonal gammopathy (indicated by arrow), in comparison with normal-pattern gamma-globulin pattern (denoted by broken line).

COMPARATIVE MEDICINE CONSIDERATIONS

Systemic lupus erythematosus is common among autoimmune diseases in the human population; reports of its prevalence vary, but 50 cases per 100,000 is a reasonable estimate. It is more common in females than in males (with a ratio of about 10:1), and it is believed that hormones may be involved in this gender disparity. In canines, SLE is rare and there is not a greater prevalence in females, most probably because the majority of female dogs are spayed (that is, undergo ovariohysterectomy), and thus lack an influence of estrogen and progesterone. The pathogenesis of and body systems involved in the disease are very similar in humans and dogs. In both species the associated skin lesions are exacerbated by exposure to ultraviolet light. Also in both humans and dogs there appears to be a genetic predisposition to development of SLE; certain HLA haplotypes show an increased incidence in humans, and there is an increased prevalence of the disease in some breeds of dog. Prednisone, a corticosteroid, is commonly used to treat SLE in both species, accompanied by other immunosuppressive agents when required.

Although it is very rare, SLE has also been described in horses and cats.

Questions

1. The urinalysis that was performed on Bover did not show any protein present in the urine. What does this tell us about the status of Bover's kidneys?

2. The plasma protein profile on Bover's bloodwork showed that he had a decreased A/G ratio. Explain what this means and how it might be reflected in the enlarged lymph nodes found on physical examination.

3. What is an ANA titer and how is the test performed?

Further Reading

Gershwin LJ (2010) Autoimmune diseases in small animals. *Vet Clin North Am Small Anim Pract* 40:439–457.

Gurevitz SL, Snyder JA, Wessel EK et al. (2013) Systemic lupus erythematosus: a review of the disease and treatment options. *Consult Pharm* 28:110–121.

Wilbe M, Jokinen P, Truvé K et al. (2010) Genome-wide association mapping identifies multiple loci for a canine SLE-related disease complex. *Nat Genet* 42:250–254.

CASE 24
PURPURA HEMORRHAGICA

EQUINE

When antigen is introduced into the body, it is recognized by the immune system as foreign, and an antibody response is initiated. Normally these antibodies are protective, and if the antigen is part of a pathogen, the antibodies facilitate elimination and/or inactivation of the pathogen. However, when an antigen persists in the body, and antibody has been produced which reacts with this antigen, the immune complexes that they create can accumulate in blood vessels and an immune mechanism called type III hypersensitivity can occur, leading to inflammation of the blood vessels (vasculitis).

Normally there are mechanisms for the safe removal of immune complexes from the body; these include binding of small complexes to erythrocytes and platelets, and removal of very large complexes by macrophages. The size of immune complexes varies depending on the ratio of antigen to antibody; those complexes with an excess of either antigen or antibody are soluble, whereas those with an antigen:antibody ratio closer to 1:1 are larger and become insoluble (Figure 24.1). The latter are readily removed by the fixed macrophages of the reticulo-endothelial system. The critical event in pathogenesis of a type III hypersensitivity response is the circulation of soluble antigen in the presence of a vigorous antibody response, which creates medium-sized immune complexes that are not readily removed by macrophages or by binding to erythrocytes and platelets, and thus circulate in the blood vascular system until they are deposited in endothelial basement membranes of capillaries and small venules. These complexes fix complement, causing production of chemotactic complement components (C3a and C5a), which attract neutrophils to the area. C3a and C5a are also called anaphylatoxins because they are able to cause mast cells to degranulate. Mast-cell mediators, such as histamine,

TOPICS BEARING ON THIS CASE:

Immune response to *Streptococcus equi*

Type III hypersensitivity; immune-complex-mediated vasculitis

M-protein antigen of *Streptococcus equi*

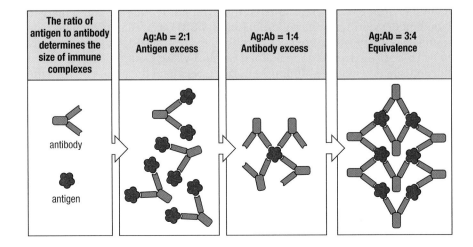

The ratio of antigen to antibody determines the size of immune complexes	Ag:Ab = 2:1 Antigen excess	Ag:Ab = 1:4 Antibody excess	Ag:Ab = 3:4 Equivalence
antibody antigen			

Figure 24.1 The ratio of antigen (shown in red) to antibody (shown in blue) determines the size of immune complexes. Antigen or antibody excess produces small soluble complexes. A ratio close to equivalence can create large insoluble complexes, which are easily removed by macrophages. Intermediate-size complexes are most likely to cause vasculitis.

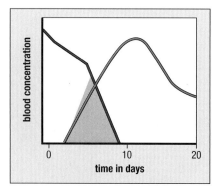

Figure 24.2 Profile of antigen and antibody concentrations that cause classical serum sickness (immune-complex-induced vasculitis). In this scenario, antigen is introduced by the intravenous route (red line) and equilibrates in the blood volume. The antigen concentration falls gradually due to metabolic degradation. When an acquired immune response (antibody) is initiated (blue line), the concentration of the antigen drops precipitously due to the formation of immune complexes (shaded area). (Adapted from Tizard IR [2008] Veterinary Immunology: An Introduction, 8th ed. Saunders.)

contribute to formation of edema. The neutrophils attracted to the site of immune-complex deposition release destructive lysosomal enzymes (including lysozyme, proteases, and collagenase), resulting in tissue damage. The blood vessels become leaky, causing local edema and hemorrhage, and some vessels become obstructed, causing tissue anoxia. The histopathology shows a leukocytoclastic vasculitis.

Type III hypersensitivity occurs in skin, joints, kidney, skeletal muscle, and other areas in which there are small blood vessels. It was first discovered when people were treated with large doses of horse serum containing passive immunity for tetanus. In this type of treatment, a bolus of foreign protein is injected into an individual, and the antibody response to that protein occurs while there is still a large amount of circulating protein remaining. The resultant disease is called "serum sickness" (Figure 24.2).

A type III hypersensitivity response can be induced by bacterial and viral diseases, as well as by self antigens in the autoimmune disease systemic lupus erythematosus (see Case 23). One bacterium that is commonly associated with a type III hypersensitivity response in horses is *Streptococcus equi*. *S. equi* typically infects young adult and sometimes geriatric horses. The disease begins with a high fever, followed by mucopurulent discharge from the nose, depression, and enlargement of the lymph nodes associated with the head. If the lymph nodes in the retropharyngeal area are greatly enlarged, this may interfere with breathing and swallowing. The lymph nodes eventually rupture and spread purulent discharge into the environment. This disease is highly contagious. Most horses recover uneventfully; however, a small percentage of equines develop a type III hypersensitivity response called purpura hemorrhagica (PH).

THE CASE OF COUNTRY GIRL: A FILLY WHO DEVELOPED VASCULAR INFLAMMATION WITH EDEMA AND MUCOSAL HEMORRHAGE

SIGNALMENT/CASE HISTORY

Country Girl is a 3-year-old Quarter Horse filly. She was sent to a training barn to begin her career as a cutting horse. In her new home, Country Girl shared a water trough with three other young horses. Approximately 1 week after arrival at the new barn she became anorectic, depressed, and had a fever of 104°F. Two days later she had a profuse mucopurulent nasal discharge and her mandibular lymph nodes were enlarged (Figure 24.3). The veterinarian was called, and Country Girl was diagnosed with "strangles" (the common name for an infection with *Streptococcus equi*). The trainer reported that several other young horses had suffered from respiratory disease with enlarged lymph nodes during the past month. Country Girl received penicillin and the trainer was instructed to hot pack the abscessed lymph node until it was ready to be lanced. The lymph node drained within 1 week; Country Girl recovered her appetite and her fever diminished. Two weeks later the trainer noticed the presence of hives (urticaria), which she treated by hosing with cold water. The next day Country Girl's legs were stocked up, her face was swollen, and she was depressed again. The vet was called back to re-examine her.

PHYSICAL EXAMINATION

On examination, Country Girl appeared to be depressed and she had an elevated temperature of 103°F (normal range, 100–102°F); her heart rate and respiratory rate were normal, without evidence of dyspnea. She walked stiffly and reluctantly, and her left rear leg was swollen with cellulitis (her right hind leg appeared normal). Ventral edema was also present. The area of the strangles

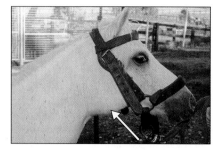

Figure 24.3 Country Girl with a *Streptococcus equi* (strangles) abscess present in the submandibular lymph node. (Courtesy of Derek Knottenbelt.)

abscess appeared to be healing, with only a slight discharge and crusts present. The ocular and oral mucous membranes showed ecchymotic hemorrhages (Figure 24.4).

DIFFERENTIAL DIAGNOSIS

The differential diagnosis for Country Girl's condition included all possible causes of vasculitis. Purpura hemorrhagica (post strangles) was most likely, because the horse had recently had a *Streptococcus equi* abscess. Other possible causes of vasculitis included *Anaplasma phagocytophilum* infection, equine infectious anemia, equine viral arteritis, equine herpesvirus infection, and immune-mediated thrombocytopenia.

DIAGNOSTIC TESTS AND RESULTS

A complete blood count (CBC) showed a mild leukocytosis and neutrophilia, which was the only abnormality detected. No organisms were seen in a buffy coat smear. An equine chemistry panel showed increased globulins and creatinine kinase. A Coggins test was negative for antibodies to equine infectious anemia virus. A culture of the draining lymph node was positive for *Streptococcus equi*. Biopsy of affected skin showed leukocytic vasculitis with evidence of edema in the surrounding tissue, and immunofluorescence of a frozen section revealed deposition of IgG in the walls of the venules.

DIAGNOSIS

Country Girl was diagnosed with purpura hemorrhagica due to the history of a recent *Streptococcus equi* infection together with the clinical signs of immune-complex disease and the absence of evidence for alternative causes.

TREATMENT

The goals of treatment for purpura hemorrhagica are to reduce the antigen load by eliminating the infection, to suppress the excessive immune response, and to support the patient as required. Country Girl was started on rifampin and procaine penicillin G to decrease the bacterial antigen load, and was then switched to oral trimethoprim sulfa. Dexamethasone was administered to suppress the immune response. Hydrotherapy of her legs was performed every 6 hours during the day. Leg wraps were used, and protective dressings were applied to open wounds. Country Girl did not have dysphagia (problems with swallowing), and so did not require feeding with a nasogastric tube. Her condition improved daily, and her dose of dexamethasone was tapered over several days and then replaced with oral prednisolone. After 3 weeks the prednisolone was stopped (after tapering), and Country Girl had become bright and energetic once again. The leg swelling diminished, and she was gradually returned to light exercise.

PURPURA HEMORRHAGICA

In horses that develop the bacterial disease strangles from infection with *Streptococcus equi*, a type III hypersensitivity (purpura hemorrhagica) is observed in less than 20% of cases. Those animals that are affected are usually either young adults or geriatric horses. The clinical signs can include depression, anorexia, ventral edema, urticaria, hind limb edema/cellulitis, muscle stiffness, ecchymotic hemorrhages on mucosal surfaces, epistaxis (rarely), and (in the more severe cases) head and neck edema that can interfere with breathing and swallowing.

There are several strangles vaccines on the market, including modified live bacteria, intranasal, bacterin (killed bacteria), and M-protein subunit. Vaccination has been associated in some cases with the development of purpura hemorrhagica, particularly when young foals are vaccinated. Vaccination

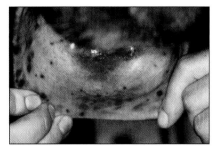

Figure 24.4 Ecchymotic hemorrhages on the oral mucosa as a result of purpura hemorrhagica. (Courtesy of Derek Knottenbelt.)

Figure 24.5 M-protein antigen is located on the fibrils of the *Streptococcus equi* bacteria. The M protein is the external *S. equi* antigen that has been shown to be involved in immune-complex formation.

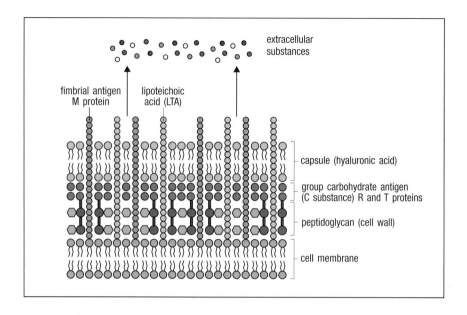

in the face of an outbreak is considered to be risky because the incidence of purpura hemorrhagica is often increased in this situation. Nonetheless, vaccination of susceptible young horses is recommended.

A surface protein of *Streptococcus*—M protein—is an antiphagocytic virulence factor involved in the formation of immune complexes with IgG and IgA in this disease. It is a surface antigen on fibrils of the bacterial cell (Figure 24.5). M protein is an important virulence factor because it binds to factor H of the alternative complement pathway and destroys the function of C3 convertase, thereby preventing the formation of C3b. Without C3b as an opsonin, phagocytosis is impaired early in infection when the alternative pathway of the complement system is most needed. Later the acquired humoral immune response facilitates recovery from the infection, but the M protein remains and binds to the antibodies produced against it, forming immune complexes that can lead to purpura hemorrhagica.

Although purpura hemorrhagica in horses is most frequently associated with infection by *Streptococcus equi*, the syndrome can occur with other diseases in which there is an antigen excess with circulating antibodies present. The pathogenesis of purpura hemorrhagica also involves similar immune mechanisms to that of systemic lupus erythematosus (SLE). Both are immune-complex-mediated type III hypersensitivity disorders; however, in SLE the antigen is a self antigen, whereas the antigen in purpura hemorrhagica is a bacterial component. Both disorders involve antibody formation and interaction with complement components that cause widespread vasculitis.

COMPARATIVE MEDICINE CONSIDERATIONS

The Henoch–Schönlein purpura (HSP) syndrome in humans is similar to purpura hemorrhagica in horses. Just as young horses are more likely to develop purpura hemorrhagica, children have the highest incidence of HSP, which is also a systemic vasculitis caused by immune-complex deposition. The antibodies involved in the immune complexes are of the IgA isotype. Immunofluorescence of skin-lesion biopsies shows IgA and C3 deposition in small blood vessels. Immune complexes are deposited in small blood vessels (venules, arterioles, and capillaries) throughout the body, but the skin and kidneys are most severely affected. Joint involvement is also common. Another similarity is that HSP usually follows an infection, including streptococcus group A. Children affected with HSP often have severe signs of colic, which is not a common feature of purpura hemorrhagica in horses.

Questions

1. What causes the edema and the ecchymosis associated with purpura hemorrhagica in affected horses?

2. Purpura hemorrhagica following a strangles infection (and sometimes after a strangles vaccine) occurs in only a small percentage of horses. Recovery from strangles and the vaccine response both involve development of an antibody response. What is different about horses that develop purpura hemorrhagica?

3. Compare purpura hemorrhagica preceded by a strangles infection with the classic example of a type III hypersensitivity response, namely serum sickness.

Further Reading

Chen O, Zhu X, Ren P et al. (2013) Henoch Schonlein Purpura in children: clinical analysis of 120 cases. *Afr Health Sci* 13:94–99.

Galan JE & Timoney JF (1985) Immune complexes in purpura hemorrhagica of the horse contain IgA and M antigen of *Streptococcus equi*. *J Immunol* 135:3134–3137.

Pusterla N, Watson JL, Affolter VK et al. (2003) Purpura haemorrhagica in 53 horses. *Vet Rec* 153:118–121.

CASE 25
CONTACT DERMATITIS

When the immune response is directed against an antigen that is not part of a pathogenic organism, the resultant inflammatory response is considered to be a hypersensitivity reaction. A hypersensitivity reaction mediated by sensitized T lymphocytes is called type IV or delayed-type hypersensitivity (DTH), and is distinct from types I, II, and III, all of which are mediated by antibodies (DTH responses are antibody independent). Immediate-type hypersensitivity (type I) is caused by IgE antibodies, and clinical signs occur within minutes after contact of a sensitized animal with the allergen. At the other end of the spectrum, a DTH (type IV) reaction takes 48–72 hours to develop after antigen contact. When a DTH response occurs, T-helper type 1 cells release cytokines, attract macrophages, and cause tissue damage. Alternatively, cytotoxic T lymphocytes can directly attack cells of the body (such as epithelium) that have the foreign antigen displayed on their surface. Both methods of damage are illustrated in Figure 25.1.

In humans, a DTH response is used to diagnose infection with some microorganisms, such as *Mycobacterium tuberculosis*. The intradermal test involves injection of a purified protein extract from *M. tuberculosis* into the skin. After 72 hours the area is observed for an erythematous indurated lesion that indicates recruitment of mononuclear cells to the area, which will only occur in a sensitized individual. The same type of response sometimes occurs naturally (in other words, not by injection with a syringe) via exposure to a contact-sensitizing agent that is able to penetrate and bind to intact skin. These small molecules, called haptens, are not able to induce an immune response on their own, but when attached to a larger protein or cell they become antigenic and stimulate T lymphocytes. The most common type of substance that can act as a hapten is a small chemical group or a metal ion. For example, chemicals incorporated into fabrics, carpets, plant resins, and metal salts can sometimes attach to host cells and become antigenic.

THE CASE OF DOMINO: A DOG WHO DEVELOPED ITCHY RED PATCHES ON HER BELLY

SIGNALMENT/CASE HISTORY

Domino is a 5-year-old spayed female mixed-breed dog. She has lived in the same house with access to a large backyard since she was purchased as a puppy by her owner. Her health has been good, and her vaccinations and flea and heartworm preventative medication are all up to date. Domino was presented to the veterinarian because over the past month she had developed

TOPICS BEARING ON THIS CASE:

Delayed-type hypersensitivity (type IV hypersensitivity)

T-helper type 1 cells

Cytotoxic T cells

Hapten

Type IV immune-mediated tissue damage		
Immune reactant	T cells	
Antigen	soluble antigen	cell-associated antigen
Effector mechanism	macrophage activation	cytotoxicity
	Th1 ⇩ cytotoxins ⇩⇩⇩	CTL ⇩
Example of hyper-sensitivity reaction	allergic contact dermatitis, graft rejection	
	rheumatoid arthritis	diabetes mellitus

Figure 25.1 In a type IV response, sensitizing substances bind to epithelial cells and, acting as haptens, cause sensitization of T lymphocytes. Activated T lymphocytes and macrophages then return to the epithelial site and cause apoptosis of epithelial cells with bound hapten, resulting in the formation of firm erythematous lesions. (From Geha R & Notarangelo L [2016] Case Studies in Immunology, 7th ed. Garland Science.)

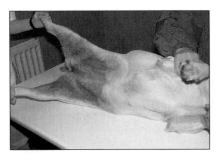

Figure 25.2 Domino showing raised erythematous lesions on the inner aspects of both femoral portions of the hind legs and on the caudal aspect of the ventral abdomen. These areas with sparse hair are likely points of contact with sensitizing substances. (From Gross TL, Ihrke PJ, Walder EJ & Affolter VK [2005] Skin Diseases of the Dog and Cat, 2nd ed. Courtesy of CRC Press.)

pruritic erythematous lesions on her ventral abdomen and on the inner aspect of her thighs. There had not been any change in diet, but some new plants were recently introduced into the yard, and weeds were recently treated with a herbicide. Upon questioning the owners, it became apparent that Domino spent many hours sleeping in a shady spot in the garden, mostly in ventral recumbency.

PHYSICAL EXAMINATION

On examination, Domino was friendly, bright, and alert, and her body condition was good, with a score of 6 out of 10.[1] Her body temperature was normal at 101.3°F, and no abnormalities were noted other than extensive erythematous lesions on the inner aspect of both her hind legs and the lower ventral midline (Figure 25.2). The lesions consisted of vesicles and excoriations from self-trauma. No flea dirt was present, and the owner confirmed regular use of Frontline® as a flea preventative. A skin scraping was negative for the presence of mites. The dog's ears were clean and not inflamed.

DIFFERENTIAL DIAGNOSIS

The location of the lesions, absence of external parasites (fleas, lice, and mites), and the history of a recent introduction of new plants into the garden and the application of herbicides (which could be the source of a chemical hapten) suggest that allergic contact dermatitis is the most likely cause of the lesions. However, irritant contact dermatitis, which is a non-allergic condition, should be ruled out. Irritant contact dermatitis is an inflammation that lacks a specific immunological cause; the lesions are a general physical reaction to caustic materials, rather than a defined immune response to a specific hapten. The pattern of skin lesions in this case does not suggest skin disease from a systemic cause such as a thyroid or corticosteroid deficiency. Atopic skin disease would show a more generalized distribution of lesions. In addition, the lesions in contact dermatitis show erythema and induration rather than fluid-filled vesicles from IgE mediated mast-cell degranulation, which would be more typical of a type I hypersensitivity/atopic dermatitis.

DIAGNOSTIC TESTS AND RESULTS

The lack of known exposure to any generally irritating chemicals made the nonspecific irritant cause of the dermatitis less likely. A diagnosis of contact dermatitis is best confirmed using a patch test in which the chemical that is suspected of acting as a hapten is placed on a gauze pad and taped onto normal skin. After 72 hours the area will show signs of erythema if the patient is sensitive to the chemical. In cases where such testing is not possible, elimination of the suspected hapten from the environment is often diagnostic. In this case, the herbicide that had been used in the garden was placed on a gauze pad and placed on a shaved spot of normal skin on the dog. Domino was also bathed to remove any residual plant resin, and was confined to the house to prevent contact with plants in the backyard. An Elizabethan (E) collar was used to prevent further self-trauma.

DIAGNOSIS

After several days the lesions began to resolve, and by 72 hours after application the skin on which the patch test was applied developed an erythematous indurated lesion. The herbicide was found to contain the hapten responsible for eliciting a contact dermatitis response.

---

[1] According to the Purina® body condition rating system.

TREATMENT

Corticosteroids (for example, prednisone) can be used to reduce the inflammatory response, but removal of the source of the contact allergen is the most critical step in sustaining resolution of the lesions caused by a contact dermatitis reaction. In this case, the owner stopped using the herbicide responsible for the response. It was also recommended that the owner should treat any secondary bacterial infection with topical ointments, and continue using an E collar to prevent licking of lesions until Domino's skin had completely healed.

CONTACT DERMATITIS

Allergic contact dermatitis is a type IV (delayed-type) hypersensitivity disorder in which a substance, usually a small chemical group, binds to host proteins or cells and elicits a T-cell-mediated response. It can be caused by chemicals found in soaps, flea collars, shampoos, wool and synthetic fibers, leather, plastic and rubber dishes, grasses and pollens, insecticides, herbicides, petrolatum, paint, carpet dyes, rubber and wood preservatives—and even the antibiotic ointment neomycin.

As shown in Figure 25.1, the induction of a strong T-helper type 1 response against a hapten causes the production of a variety of mediators and chemokines that activate the affected epithelium macrophages and additional lymphocytes. For this to occur, the hapten binds to endogenous proteins and is processed by Langerhans cells in the skin. Presentation by major histocompatibility complex (MHC) class II molecules of the altered peptides to T cells results in the activation of T-helper type 1 cells. The inflammatory cells accumulate over the course of several days, and macrophages activated by interferon-γ contribute further to the inflammation by releasing additional mediators and causing epithelial cells to secrete pro-inflammatory cytokines. The disease is present only at sites of contact with the allergen, where epithelial cells that have bound allergenic haptens are attacked by the immune response.

Some chemicals, such as plant oils like those from poison ivy, are lipid soluble and cross the cell membrane. Once in the cytosol they bind to peptides, and these modified peptides are presented by MHC class I molecules to cytotoxic CD8$^+$ T cells (also known as cytotoxic T lymphocytes, or CTLs). These CTLs are capable of directly killing epithelial cells that have bound to the chemical.

COMPARATIVE MEDICINE CONSIDERATIONS

Many people who have had extensive exposure to nickel in jewelry (for example, earrings) and watch straps and buckles develop contact hypersensitivity to the nickel. The hapten in these cases is nickel sulfate, $NiSO_4(H_2O)_6$, which is formed when perspiration comes into contact with the nickel. The development of allergic contact dermatitis in response to plant resins is also common in humans. The plant resin urushiol is present in many plants, but is particularly concentrated in poison oak, poison ivy, and poison sumac. When the resin comes into contact with skin, either directly from a plant or via a towel or clothing, it binds to the skin and acts as a hapten. The immune response that causes the contact allergy is directed towards a complex of urushiol derivatives, called pentadecacatechol, which is bound to skin proteins. The epithelial cells are recognized as foreign by host lymphocytes, and an immune attack is initiated.

Horses can develop allergic contact dermatitis in response to a variety of haptens found on tack, stable blankets and pads, residues from tack cleaning, and insecticides and other sprays. Cats have also been shown to develop contact allergy to plant resins, topical medications, and wool.

Questions

1. How does the contact or delayed-type hypersensitivity response differ from an immediate-type hypersensitivity reaction with regard to initiating immune reactant, physical appearance of the lesions, and timing of the appearance of lesions?

2. How do plant chemicals such as urushiol oil cause skin lesions from allergic contact dermatitis?

3. Would testing blood for antibodies to a sensitizing chemical be a helpful diagnostic tool in this case, or not? What is the reason for this?

Further Reading

Comer KM (1988) Carpet deodorizer as a contact allergen in a dog. *J Am Vet Med Assoc* 193:1553–1554.

Murayama N, Midorikawa K & Nagata M (2008) A case of superficial suppurative necrolytic dermatitis of miniature schnauzers with identification of a causative agent using patch testing. *Vet Dermatol* 19:395–399.

Olivry T, Prélaud P, Héripret D & Atlee BA (1990) Allergic contact dermatitis in the dog. Principles and diagnosis. *Vet Clin North Am Small Anim Pract* 20:1443–1456.

Thomsen MK & Kristensen F (1986) Contact dermatitis in the dog. A review and a clinical study. *Nord Vet Med* 38:129–147.

CASE 26
AUTOIMMUNE
THYROIDITIS

Autoimmune disease occurs when the ability of the immune system to recognize self and non-self breaks down. This process can involve development of autoantibodies against self molecules or tissue antigens, as well as T cells that recognize self antigens. When an autoimmune response targets a specific tissue or organ, the clinical signs reflect the alterations caused in that tissue or organ. For example, immune responses directed against the islet cells of the pancreas will decrease insulin production and cause diabetes, and destruction of cells in the thyroid gland that produce thyroxine will cause signs relevant to low thyroid hormone levels (hypothyroidism). These organ-specific autoimmune diseases stand in contrast to systemic autoimmune disease, such as systemic lupus erythematosus (see Case 23), in which multiple body systems can be affected.

It is estimated that about 50% of the cases of hypothyroidism in dogs are caused by immune-mediated hypothyroidism. The thyroid gland is composed of follicles containing colloid and surrounded by epithelial cells. The gland selectively acquires iodine and uses it to make thyroglobulin, which is stored in the colloid of the follicles. The follicular epithelial cells secrete triiodothyronine (T3) and tetraiodothyronine (T4, also known as thyroxine)—the thyroid hormones. This process occurs under the direction of thyroid-stimulating hormone (TSH), which is produced in the pituitary gland. TSH causes the follicular epithelial cells to take up the thyroglobulin by endocytosis and then remove iodine molecules to produce T3 and T4 for secretion into the blood. Following secretion, the T4 is eventually converted to T3 after removal of one iodine molecule. These hormones are critical for many functions in the body, including metabolism.

THE CASE OF JENNY: A DOG WHO UNDERWENT A TRANSFORMATION FROM PLAYFUL AND VIGOROUS TO DULL AND LETHARGIC

SIGNALMENT/CASE HISTORY

Jenny is a 9-year-old spayed female Golden Retriever. She was purchased from a breeder as a puppy. For most of her life she has enjoyed playing ball and chase with the children of the family, and was always up for a good game of Frisbee. She lives indoors but has daily access to a large enclosed yard. There are no other dogs in the household. Jenny is fed Purina ONE® daily, and has unlimited access to fresh water. She is on a regular Frontline® flea prophylaxis program. Over the course of the past year her owners have noticed that she has

TOPICS BEARING ON THIS CASE:

Autoimmunity

Tissue destruction by autoantibodies

Tissue destruction by autoreactive cytotoxic T cells

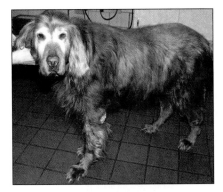

Figure 26.1 Jenny with lethargic appearance and a sparse hair coat. (Courtesy of Danny W. Scott.)

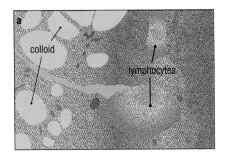

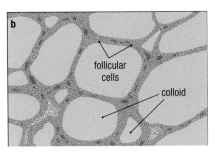

Figure 26.2 (a) Illustration of a thyroid biopsy showing few remaining normal follicles and diffuse infiltration of lymphocytes, plasma cells, and macrophages throughout the tissue. These lesions are typical of lymphocytic thyroiditis. The lymphocytes are thought to cause destruction of epithelial cells in the thyroid gland by either antibody-dependent cell-mediated cytotoxicity or CD8[+] T-cell cytotoxicity, or both. (b) Normal thyroid structure, shown for comparison.

become increasingly lethargic and she has gained weight, although no change has been made to her diet. Her once shiny hair coat has become dull, sparse, and somewhat greasy, with dandruff (Figure 26.1). She frequently scratches at her ears, which exude a musty odor. During the last month Jenny has become obtunded, with little interest in anything but sleeping.

PHYSICAL EXAMINATION

On physical examination, Jenny appeared dull and only moderately responsive. She had mild tartar on her canine and molar teeth, and slightly pale mucous membranes. Her heart and respiratory rates were within normal limits and no heart murmurs were detected. Jenny's hair coat was sparse over the thorax bilaterally, with seborrhea present. There was evidence of self-trauma to the ears, and a waxy exudate was present. No fleas were visible. Jenny was well hydrated, but her weight was excessive.[1] The musculoskeletal system appeared normal. Abdominal palpation did not reveal any enlargement of the liver or spleen. Peripheral lymph nodes were within normal limits.

DIFFERENTIAL DIAGNOSIS

The differential diagnosis for these clinical signs includes hypothyroidism/ autoimmune thyroiditis, pituitary-mediated hypothyroidism, and thyroid tumor, all of which can be ruled out by thyroid biopsy. Possible secondary problems include anemia, as indicated by pale mucous membranes, and otitis externa, which can be secondary to hypothyroidism.

DIAGNOSTIC TESTS AND RESULTS

An ultrasound guided biopsy of the thyroid was performed, and this showed that the thyroid gland was enlarged. A small punch biopsy was obtained for histopathological evaluation. Histology showed infiltration with lymphocytes, and obliteration of much of the normal follicular structure (Figure 26.2). A canine chemistry panel revealed cholesterol levels of 706 mg/dL (normal range, 135–345 mg/dL). Thyroxine (T4) levels were low at < 0.3 µg/dL (normal range, 1–3.6 µg/dL); free T4 levels were also low at 0.15 ng/µL (normal range, 0.8–3.5 ng/µL), and the TSH level was high at 2.7 ng/µL (normal range, 0–0.6 ng/µL). The test for antithyroid antibodies was positive (normal result is negative). The complete blood count showed the presence of anemia, with a low red blood cell count of 4.13×10^6 (normal range, $5.6–8 \times 10^6$), a low hemoglobin level of 10.9 g/dL (normal range, 14–19 g/dL), and a low hematocrit of 29.4% (normal range, 40–55%). A lack of reticulocytes indicated that the anemia was nonregenerative.

DIAGNOSIS

Based on the blood test results, Jenny was found to have a moderate anemia, elevated serum cholesterol levels, low T4 levels, elevated TSH levels, and the presence of autoantibodies against thyroglobulin. These findings were consistent with a diagnosis of autoimmune thyroiditis (also called Hashimoto's disease). The presence of antithyroglobulin antibodies is an important diagnostic feature of this disease.

TREATMENT

Jenny was started on thyroxine replacement therapy. Her weight at the time of examination was 80 pounds, and she was started on a dose of 0.1 mg/pound of levothyroxine (Soloxine®), to be given on an empty stomach. Her otitis was treated with the antibacterial, antimycotic, and anti-inflammatory ointment Entederm®, and she was given prescribed shampoo to treat the seborrhea.

---

[1] According to the Purina body condition rating system.

When she returned for a check-up 1 month later her owners reported that her energy levels were much improved and she had stopped scratching at her ears. At the 6-month visit she had regained her pre-symptom appearance and mentation. A thyroid panel taken at this time showed that her T3 and T4 levels were within the normal range. Jenny will have to remain on levothyroxine therapy for the rest of her life, and will need to have the levels checked several times a year to ensure that her dose remains effective.

AUTOIMMUNE THYROIDITIS

When immune tolerance to antigens of the thyroid gland breaks down (in most cases it is not known why this happens), an immune attack on the thyroid gland ensues. In the case of lymphocytic thyroiditis, there is an intense infiltration of the gland by lymphocytes (primarily T cells), and the gland is gradually destroyed. It usually takes at least a year for enough thyroid tissue (an estimated 75%) to be lost to cause sufficient depression of T3 and T4 levels and consequently the clinical signs of hypothyroidism. In addition to the cytotoxic T cells working to destroy the thyroid, autoantibodies to thyroglobulin, T3, T4, and thyroid peroxidase are being produced, and these can initiate an immune attack on the thyroid gland. These antibodies serve as a useful serological marker of immune-mediated thyroid disease.

The clinical signs of hypothyroidism include changes in mentation (dullness, sluggishness), weight gain, increased sensitivity to cold, changes in appearance (for example, drooping eyelids), changes in hair coat (including hair loss, dry, flaky, or oily skin, and hyperpigmentation of skin) and often otitis, with itching and infection of the ear canal. Some very severely affected dogs may exhibit neurological signs, such as seizures. Diagnosis depends on measurements of the levels of T3, T4, and TSH in the blood. In a dog with this disease, the T3 and T4 levels will be lower than normal, and the TSH level will be higher than normal, as the pituitary gland senses that there is insufficient T3 and T4 and releases more TSH in an attempt to stimulate T3 and T4 production. The determination of an immune etiology requires testing for antibodies to T3 and T4 and thyroglobulin. A biopsy of the thyroid gland will reveal a lymphocytic infiltrate.

It is known that lymphocytic thyroiditis has a genetic component. Certain breeds are more likely to develop it, particularly the Labrador Retriever, Golden Retriever, Cocker Spaniel, Boxer, and Doberman Pinscher, although the disease has been reported in many other breeds and also in mixed-breed dogs. A recent study has shown a significant association with DLA-DQA1*00101 in a series of 173 hypothyroid dogs of various breeds. This association with a specific dog leukocyte antigen (DLA) is not surprising, as many other autoimmune diseases of dogs are also associated with DLA, just as human autoimmune diseases are commonly associated with one or more human leukocyte antigen (HLA) types.

Several reports in the literature suggest that the interaction of environment and genetics is a major predisposing factor for many autoimmune diseases; this is also true for thyroiditis. Exposure to or infection with certain viruses may predispose to loss of self tolerance. One theory of viral intervention is that the virus engages in molecular mimicry, such that the immune response to a virus can activate a T cell to respond to antigens on the host's thyroid gland. Another possible mechanism for development of autoimmunity is nonspecific activation of bystander cells by excessive cytokine production, perhaps in response to an infection.

COMPARATIVE MEDICINE CONSIDERATIONS

Hashimoto's thyroiditis is the name given to lymphocytic (autoimmune) thyroiditis in human patients. This disease was first identified and reported in 1912 by Hakaru Hashimoto, and is the most common cause of hypothyroidism in humans. It is estimated that up to 1.5 in every 1000 people have this disease. Women are affected more frequently than men, as is true for several other autoimmune diseases. The clinical signs are similar to those in dogs, namely decreased activity, mental dullness, and sensitivity to cold. Antibodies to thyroglobulin and thyroid peroxidase are present in the blood, and the thyroid gland is infiltrated with lymphocytes. In humans there is a strong genetic association—HLA DR5 has a relative risk of 3, which means that these individuals are three times more likely to develop the disease than individuals who do not have DR5. Other factors, such as environmental exposure and viral infections, have also been implicated in development of the disease.

Questions

1. Dogs with autoimmune thyroiditis have antibodies against thyroglobulin, T3, and T4. By what immune mechanism(s) could these antibodies cause hypothyroidism?

2. Infiltration of the thyroid follicles with lymphocytes is a sign that activated T cells have a role in thyroid gland destruction. By what mechanism could these cells cause thyroid gland destruction?

3. Explain why so many body systems are affected in a patient with hypothyroidism.

Further Reading

Graham PA, Nachreiner RF, Refsal KR & Provencher-Bolliger AL (2001) Lymphocytic thyroiditis. *Vet Clin North Am Small Anim Pract* 31:915–33, vi–vii.

Kennedy LJ, Quarmby S, Happ GM et al. (2006) Association of canine hypothyroidism with a common major histocompatibility complex DLA class II allele. *Tissue Antigens* 68:82–86.

Nachreiner RF, Refsal KR, Graham PA et al. (1998) Prevalence of autoantibodies to thyroglobulin in dogs with nonthyroidal illness. *Am J Vet Res* 59:951–955.

Wilbe M, Sundberg K, Hansen IR et al. (2010) Increased genetic risk or protection for canine autoimmune lymphocytic thyroiditis in Giant Schnauzers depends on DLA class II genotype. *Tissue Antigens* 75:712–719.

CASE 27
RECURRENT AIRWAY OBSTRUCTION

EQUINE

The immune response to inhaled pathogens is an important component of protective immunity, but when this immune response is directed toward antigens in the environment—such as dust, molds, and pollens—the result is a type I or type III hypersensitivity response. In a type III response, IgG antibodies against the inhaled antigens are produced. These form immune complexes and cause complement fixation. The anaphylatoxins C3a and C5a degranulate mast cells, which release histamine and other mediators that can cause smooth muscle contraction in the airway, thereby preventing normal airflow. These same molecules are chemotactic for neutrophils, which release enzymes that cause further inflammation and damage to the lung. The neutrophilia that is seen in lungs and in lung lavage fluid from horses with recurrent airway obstruction (RAO) fits well with this pathogenesis.

In a type I response, IgE that is specific for the mold and dust antigens is detectable in blood and possibly in lung lavage fluid. IgE can also cause bronchoconstriction by mast-cell degranulation, but by a different mechanism—the IgE binds to the mast-cell Fc epsilon receptors, and triggers degranulation when antigen binds to the Fab portions of the mast-cell-bound IgE. Eosinophils are a hallmark of IgE responses, and some eosinophils would be expected in lung, respiratory secretions, and sometimes at increased levels in peripheral blood. In cases of RAO, horses sometimes have both IgG and IgE antibodies present, which suggests a mixed pathogenesis.

THE CASE OF GEOFF: A HORSE WHO EXPERIENCED EPISODES OF DIFFICULT BREATHING AND COUGHING WHEN HE WAS FED

SIGNALMENT/CASE HISTORY

Geoff is a 12-year-old Canadian Warmblood gelding. He is stabled in a stall bedded with straw, with no direct access to the outdoors. His owners turn him out into a small paddock daily for exercise, but Geoff spends the rest of the time in his stall. The owners reported that Geoff started coughing at feeding time a few years ago, and that this has now progressed to include episodes during which he seems to struggle for breath. During these times Geoff can often be heard wheezing. He has also had a mucoid nasal discharge intermittently. Geoff eats grass hay twice daily and a grain supplement with vitamins daily. His vaccinations and worming are up to date.

TOPICS BEARING ON THIS CASE:

Hypersensitivity response (types I and III)

Pathogenesis of airway obstruction

Lung inflammation

Genetic predisposition

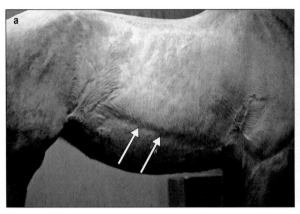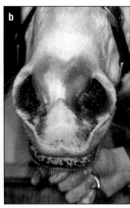

Figure 27.1 The presence of a heave line is shown in panel (a) with arrows. Panel (b) shows the flared nostrils associated with difficulty breathing. (Courtesy of Derek Knottenbelt.)

PHYSICAL EXAMINATION

On examination, Geoff appeared alert and slightly excited with a temperature of 100°F (normal range, 99–101°F), respiratory rate of 63 breaths/minute (normal range, 10–24 breaths/minute), and pulse of 40 beats/minute (normal range, 29–44 beats/minute). His mucous membranes were pink, with a capillary refill time of less than 2 seconds. His body condition was good, but a heave line was present (Figure 27.1A), indicating frequent use of abdominal muscles to assist breathing. Body systems other than the respiratory tract were within normal limits. Examination of the respiratory system showed flared nostrils (Figure 27.1B), with airflow present bilaterally. A tracheal rattle was present, and there were harsh bronchovesicular sounds in the cranioventral area of the thorax. Intermittent abdominal muscle movement was also noted. A rebreathing examination was performed by placing a soft plastic bag over the horse's muzzle while the thorax was auscultated to enhance detection of crackles and wheezes. Geoff did not tolerate the rebreathing examination well. It revealed expiratory wheezes in the ventral lung quadrants, and overall harsh lung sounds.

DIFFERENTIAL DIAGNOSIS

Recurrent airway obstruction (RAO) is most likely, taking into consideration the case history of a gradually increasing respiratory sensitivity exhibited at feeding time, and the physical examination, which did not show evidence of an infectious cause, but did demonstrate severe dyspnea and wheezing. Pneumonia should be considered, either as a single entity or secondary to RAO. Less likely would be neoplasia, either primary or metastatic to the lung.

DIAGNOSTIC TESTS AND RESULTS

Blood was obtained for a complete blood count (CBC) and chemistry panel. The results of the CBC showed moderate neutrophilia at 8029/μL (normal range, 2600–6800/μL), and an eosinophilia at 622/μL (normal range, 0–200/μL). The equine chemistry panel was essentially normal, with the exception of a slight increase in total plasma protein at 8.9 mg/dL (normal range, 5.8–7.7 mg/dL) and globulins at 5.6 mg/dL (normal range, 1.6–5 mg/dL). A tracheal wash and a bronchoalveolar lavage (BAL) were also performed; review of the cyto-centrifuged slide prepared from the fluid showed the presence of numerous non-degenerated neutrophils and mononuclear cells (Figure 27.2A). Normal BAL consists primarily of mononuclear macrophages (Figure 27.2B). Thoracic radiographs showed that a moderate diffuse bronchointerstitial pattern was

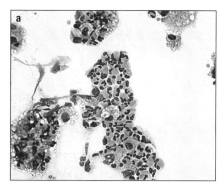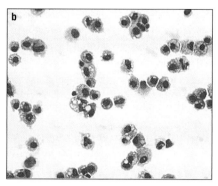

Figure 27.2 (a) Cells from the bronchoalveolar lavage, which are predominantly neutrophils, indicating an inflammatory response in the airways. (b) Cells from a normal bronchoalveolar lavage, which are predominantly mononuclear macrophages. (From Munroe GA & Weese JS [2011] Equine Clinical Medicine, Surgery, and Reproduction. Courtesy of CRC Press.)

present. There were no signs of pneumonia or nodular densities. Serum was sent to a laboratory to test for IgE antibodies against a panel of allergens (serum ELISA). The results indicated strong reactivity to several molds (*Mucor* mix, *Penicillium*, *Stemphylium*, *Alternaria*, and *Cephalosporium*) and to dust mites and storage mites.

DIAGNOSIS

A diagnosis of recurrent airway obstruction (RAO) was made based on the case history, the clinical signs, and the absence of infectious disease. The CBC and chemistry panel supported the diagnosis.

TREATMENT

Geoff was treated with dexamethasone (initially 40 mg once a day and tapered to 10 mg every other day by the end of a month), Clenbuterol 10 mL twice a day by mouth, and his hay was soaked in water to decrease the amount of dust. The owner was told to try to avoid dusty and moldy food hay and stall bedding. Administration of the inhaled steroid fluticasone (Flonase®) in an equine inhalation chamber was advised during seasonal flare-ups.

RECURRENT AIRWAY OBSTRUCTION

RAO is a common disease of stabled horses in which a hypersensitivity to inhaled organic antigens develops. Horses that are fed moldy hay or bedded on moldy straw inhale antigens which, over time, cause severe lung inflammation. The age of onset is usually 10–12 years, and the disease worsens over time. Initially a cough occurs when the horse is exposed to the antigen (often at feeding time if the antigen is in the hay). This progresses to dyspnea and a mucoid nasal discharge. As the disease worsens and the horse develops airway remodeling, bronchial constriction makes breathing more difficult, resulting in use of the abdominal muscles to express air from the lungs. These muscles hypertrophy over time and cause the development of a heave line. When a horse becomes severely dyspneic, the nostrils flare with the increased effort to breathe.

Often the antigens that are causing hypersensitivity come from infestation of hay with fungal species such as *Aspergillus fumigatus*, *Alternaria* species, and *Penicillium* species. Both dust mites and hay storage mites can also infest hay and act as antigens, and there is a seasonal form of the disease which occurs in horses that are housed on grass pasture. In all of these cases, inhalation of environmental antigens stimulates an immune response in the lung that is characterized by an influx of neutrophils to the bronchoalveolar space. The pathogenesis of this disorder is complex, because although the clinical signs resemble asthma in many ways, the role of IgE-mediated type I hypersensitivity has been debated, and IgE is detected in some cases but not in others. Serum IgG antibodies against mold antigens are detectable in most cases. These antibodies initiate a type III hypersensitivity response with induction of C-X-C cytokines and attraction of neutrophils to the lung. It has been shown that the antigens from *Aspergillus fumigatus* express pathogen-associated molecular patterns (PAMPS) that bind to receptors on dendritic cells in the lung to initiate production of pro-inflammatory cytokines. Binding of phospholipomannan from *Aspergillus fumigatus* to toll-like receptor 2 initiates production of cytokines IL-10 and TGFβ. In addition, studies have shown the induction of CD4+ CD25+ FoxP3+ T-regulatory (Treg) cells in response to inhalation of *Aspergillus*. These cells have been proposed as down-regulators of the inflammatory response. It is reported that Treg cells arriving early during the inflammatory process can decrease the influx of neutrophils to the lung, whereas those that arrive later at the lung produce the immunomodulatory cytokine IL-10, but are less capable of inhibiting the neutrophil influx. There are antigens other than mold that stimulate RAO, and there is a seasonal form of the disease which occurs in horses that are housed on grass pasture.

A genetic predisposition to the development of RAO has been described. For example, a higher than expected incidence of the condition has been documented in several families of Warmblood horses. These genetically RAO-susceptible horses have also been shown to have an increased resistance to strongylid nematodes (a type of intestinal parasite). Further studies on this relationship may enhance our understanding of the role of IgE in RAO, as IgE responses to gastrointestinal nematodes are well recognized as a protective defense mechanism.

COMPARATIVE MEDICINE CONSIDERATIONS

Although RAO has many features in common with human asthma, the pathogenesis and lung pathology are not identical. When horses with RAO have been skin tested for IgE reactions to environmental antigens (for example, molds), some test positive, but there is a high incidence of similarly positive reactions in clinically healthy horses. This means that the role of IgE in the pathogenesis of RAO is still somewhat ambiguous, whereas the role of IgE in human asthma is paramount.

Humans can develop hypersensitivity pneumonitis to a variety of inhaled organic antigens, including *Aspergillus fumigatus*. The antigen involved stimulates an IgG response and a resultant type III hypersensitivity reaction in the lungs. A common example is farmer's lung disease (extrinsic allergic alveolitis), which occurs when the human feeder inhales mold spores from hay; interestingly, the same condition occurs in cattle that are fed the moldy hay. IgG antibodies that react with *Saccharopolyspora rectivirgula* (formerly *Micropolyspora faeni*) or other thermophilic actinomycetes are found in the serum both of affected cattle and of their handlers. Affected individuals have decreased lung compliance. Neutrophils infiltrate the lung, stimulated by the release of chemokines and pro-inflammatory cytokines. Many features of this disease are similar to those of RAO in horses.

There are references to RAO in the literature in which the descriptive term "chronic obstructive pulmonary disease (COPD)" is used. Lung inflammation and the effect of RAO on breathing have similarities to COPD in humans. However, these are very different diseases. COPD is defined as several clinical syndromes that obstruct breathing, including chronic bronchitis and emphysema, and is associated with the smoking of cigarettes. The cigarette smoke causes a neutrophil influx with release of elastase, which together with the production of matrix metalloproteases by macrophages induces tissue damage.

Questions

1. It is very likely that the pathogenesis of RAO involves a type III and/or a type I hypersensitivity response. Briefly describe the components of each of these two types of immune response, and explain how both could lead to a clinical syndrome in which a horse develops airway obstruction in response to antigen inhalation.

2. There is some evidence of a genetic predisposition to RAO. Is this likely to involve a type I hypersensitivity or not? What are the reasons for this?

3. What is the reasoning for treating the case described above with dexamethasone?

Further Reading

Bründler P, Frey CF, Gottstein B et al. (2011) Lower shedding of strongylid eggs by Warmblood horses with recurrent airway obstruction compared to unrelated healthy horses. *Vet J* 190:e12–e15.

Davis E & Rush BR (2002) Equine recurrent airway obstruction: pathogenesis, diagnosis, and patient management. *Vet Clin North Am Equine Pract* 18:453–467, vi.

Henríquez C, Perez B, Morales N et al. (2014) Participation of T regulatory cells in equine recurrent airway obstruction. *Vet Immunol Immunopathol* 158:128–134.

Lanz S, Gerber V, Marti E et al. (2013) Effect of hay dust extract and cyathostomin antigen stimulation on cytokine expression by PBMC in horses with recurrent airway obstruction. *Vet Immunol Immunopathol* 155:229–237.

Tahon L, Baselgia S, Gerber V et al. (2009) *In vitro* allergy tests compared to intradermal testing in horses with recurrent airway obstruction. *Vet Immunol Immunopathol* 127:85–93.

CASE 28
FLEA ALLERGY
DERMATITIS

Skin disease is one of the most common afflictions of the domestic dog. Itchy skin (pruritus) leads to scratching, with trauma to the skin that often results in secondary bacterial or yeast infection. Ectoparasites—particularly fleas, lice, and mites—are a common cause of canine skin disease. In most cases a well-performed skin scraping and an examination of the skin for the presence of "flea dirt" will facilitate diagnosis of mites and fleas, respectively. Once it has been diagnosed and treated, in most dogs a flea infestation can be eliminated and with it the itch–scratch cycle. However, for some dogs fleas lead to a far more severe problem that requires vigilant prophylaxis and the use of anti-inflammatory drugs to facilitate skin healing and cessation of clinical signs. When a flea bites a dog it injects a small amount of its salivary proteins into the skin. These salivary proteins can be processed by Langerhans cells in the skin and induce an immune response. In some dogs, production of IgE antibodies against the flea salivary antigens facilitates sensitization of mast cells and subsequent degranulation, with release of vasoactive mediators, chemokines, and an influx of eosinophils. For these dogs, the bite of a single flea can be sufficient to initiate the allergic response.

TOPICS BEARING ON THIS CASE:

Type I hypersensitivity

Type IV hypersensitivity

Immune response to ectoparasites

THE CASE OF WALDORF: A DOG WITH CHRONIC ITCHY SKIN

SIGNALMENT/CASE HISTORY

Waldorf is a 4-year-old castrated male Irish Setter with a body weight of 35 pounds, living in a temperate climate. He spent the first 3 years of his life with a family who took excellent care of him, treating him with a flea prophylactic (Frontline®) monthly. He had no skin problems. His fourth year has been spent with a new family whose household includes another dog and several cats, all of which have access to a large backyard. Waldorf's new owners have not maintained the monthly Frontline® regimen. For the past few months he has been scratching repeatedly, biting at his hind end and rubbing on furniture to alleviate the itching. A reddened hairless area has formed on his back just in front of his tail (Figure 28.1). In answer to the veterinarian's questioning about the flea status of the other animals in the household, Waldorf's owners admitted that they were not aware of any fleas, but that all of the animals scratched themselves, although none of them scratched as much as Waldorf.

PHYSICAL EXAMINATION

On examination, Waldorf was bright, alert, and well hydrated. He tried to scratch several times while standing on the examination table. All body systems were within normal limits, with the exception of mild dental tartar,

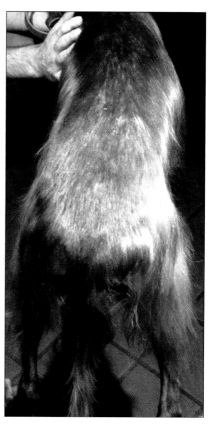

Figure 28.1 Alopecia and erythema on Waldorf's dorsal lumbosacral area, cranial to the head of the tail. Close examination would be likely to reveal some flea dirt. (Courtesy of Danny W. Scott.)

slightly enlarged popliteal lymph nodes, and an area of erythema and alopecia over his lumbosacral area and the cranial dorsal portion of his tail. On closer examination, several fleas were observed, as well as a moderate amount of reddish-black material ("flea dirt").

DIFFERENTIAL DIAGNOSIS

The differential diagnosis for a highly pruritic skin condition associated with alopecia includes atopic dermatitis, mite infestation (species of the genera *Sarcoptes* and *Demodex*), and flea allergy dermatitis (FAD).

DIAGNOSTIC TESTS AND RESULTS

The observation of fleas and flea dirt suggested the possibility of flea allergy. Serum IgE testing was performed to determine whether there was a type I hypersensitivity allergic component to Waldorf's flea problem, and to test for reactivity to other allergens that might be the cause of atopic dermatitis. The serum IgE results showed positive reactivity to both flea salivary allergens and whole flea antigen. Skin scrapings were performed to rule out other causes of the pruritic skin condition. The scrapings were negative for *Demodex*, *Sarcoptes*, and other mites. A fecal sample was tested for the presence of the parasitic worm *Dipylidium caninum*, and was positive.

DIAGNOSIS

A diagnosis of flea allergy dermatitis (FAD) is often made on the basis of history and physical examination. The presence of fleas and the absence of other ectoparasites, with a typical distribution of inflammatory lesions, provides strong suggestive evidence of FAD. The serum IgE testing that demonstrated IgE reactivity to flea allergens confirmed this diagnosis for Waldorf.

TREATMENT

Waldorf was started on a regimen of topically applied flea preventative medication. His owners were advised that there are several excellent products available that could be used to keep Waldorf flea-free. It was also recommended that all the cats and the other dog in the household should be treated for fleas and maintained on monthly prophylactic therapy. Since the fleas jump off the dog to lay their eggs, the owners were also advised to treat all bedding used by the family pets, and to utilize foggers in the indoor and outdoor areas that they frequented. To bring the inflammation under control, Waldorf was placed on a course of corticosteroid therapy, consisting of prednisone 1.5 mg/kg divided into two daily doses for 3 days, then 1 mg/kg for another 3 days, and finally 0.5 mg/kg for an additional 10 days. Thereafter he received 0.5 mg/kg every other day for another 2 weeks. He was also treated with a parasiticide to eliminate his tapeworm infection. Dogs with fleas often have tapeworms (*Dipylidium caninum*) because, while biting at their skin, they ingest fleas that are the intermediate host of the tapeworm.

FLEA ALLERGY DERMATITIS

Flea allergy dermatitis is a type I and/or type IV hypersensitivity disease of the skin, in which the canine patient produces IgE antibodies against salivary antigens of the flea (Figure 28.2). This hypersensitivity usually develops in young adult dogs. The fleas that commonly infest dogs and cats are *Ctenocephalides canis* and *Ctenocephalides felis*. These ectoparasites take blood meals by biting into the dermis, which has a rich supply of blood vessels. The flea's saliva contains many proteins, some of which are potent enzymes in the size range 18–32 kDa—the ideal size for antigenicity. Dendritic cells of the skin (Langerhans cells) take up these flea antigens, process them, and move with them to the local lymph node where they are presented on major histocompatibility complex

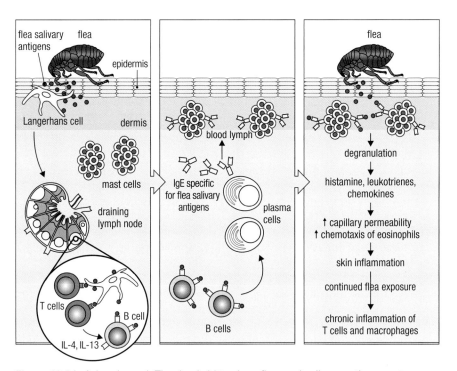

Figure 28.2 Left-hand panel: The dog is bitten by a flea, and salivary antigens enter through the epidermis, where Langerhans cells pick them up for antigen processing. The Langerhans cells enter a lymphatic vessel and travel to the draining lymph node. During this process they up-regulate major histocompatibility class II (MHC II) molecules. Once in the lymph node they find T lymphocytes in lymphoid follicles and present the antigenic peptides on MHC II molecules to T cells with receptors specific for the antigenic peptide being presented. B cells in lymphoid follicles with specific receptors bind antigen and receive co-stimulatory signals from T cells (including IL-4 and IL-13). Central panel: B cells differentiate into plasma cells that make IgE specific for the flea salivary antigens. The IgE enters the circulation and binds to mast cells in the dermis via high-affinity IgE receptors. At this point the dog is sensitized to flea salivary antigens. Right-hand panel: Subsequent exposure of the dog to flea saliva will cause the mast cells to release preformed mediators as a result of IgE cross-linking by antigens entering through the flea bite. The formation of leukotrienes and synthesis of cytokines and chemokines are stimulated. The immediate effects of histamine and the later effects of the leukotrienes and chemokines cause clinical signs of inflammation and pruritus. If allergen exposure is chronic, additional mechanisms such as T-cell-mediated (type IV) inflammation may occur, creating chronic skin lesions.

(MHC) class II molecules to the follicular T-helper type 2 cells. Antigen also enters the lymph and travels to the lymph nodes, where it binds to receptors on B lymphocytes. These signals initiate the production of cytokines IL-4 and IL-13, which further stimulate the B cells to differentiate into plasma cells that produce IgE. Those dogs that are atopic are most likely to respond to the insult by producing IgE antibodies, which bind to dermal mast cells.

The exquisite sensitivity of the type I hypersensitivity response explains how the bite of only one or two fleas can trigger such a rapid pruritic response. Indeed, a hallmark of flea allergy dermatitis compared with simple flea infestation is the severity of the response in the absence of a large number of fleas. The release of mediators such as histamine and the stimulation of eicosanoid production create intense pruritus, which stimulates biting and scratching, with accompanying hair loss and erythema. Exposure of a flea-allergic dog to an intradermal injection of flea antigen will usually result in the development of an erythematous wheal within 15 minutes. Those dogs that have a delayed-type hypersensitivity response to flea allergens will usually show an erythematous indurated lesion within 24–48 hours. Some dogs may exhibit both types of immune response during the course of the allergy.

The initial lesions from the flea bite are erythematous plaques and wheals, but self-trauma can cause hypotrichosis, ulcers, and erosions. The resulting excoriation of the skin often creates an optimal environment for secondary infection by bacteria and yeasts such as *Staphylococcus intermedius* and *Malassezia pachydermatis*. Thus secondary pyoderma is common. If the dog remains untreated and flea infestation continues, chronic changes can occur, such as lichenification, crusting, alopecia, and hyperpigmentation. In these cases the lesions may remain confined to the dorsolumbar area, or they may spread to other body areas.

Although the initial immune response to flea salivary antigens is usually an IgE-mediated hypersensitivity, chronic cases of flea allergy dermatitis (FAD) are sometimes associated with a type IV hypersensitivity. The type IV reaction is a delayed-type hypersensitivity, which is mediated by sensitized $CD4^+$ T-helper type 1 lymphocytes. These cells reside in the epithelial area that contains the antigen, and release cytokines IL-12 and interferon γ, which stimulate T cells and macrophages to accumulate within the epithelium. Accumulation of lymphocytes and macrophages is characteristic of this type of response, and accounts for the indurated nature of the lesion. Biopsy of affected skin is the most effective method of differentiating type I from type IV (delayed) hypersensitivity. There are reports of variable results of flea IgE serum testing, with one study showing up to 88% accuracy in differentiating between flea-allergic and flea-naive dogs.

Immunotherapy for FAD has been investigated, but this therapy is not commonly used in clinical practice. Vaccines consisting of purified flea salivary allergen or whole flea allergen have been tested in dogs with FAD, and the results have not shown any difference between vaccinated and control groups. In other studies, a vaccine was developed using flea hindgut antigen, with the intention that dogs would develop antibodies to these antigens, which would bind the flea hindgut cells and initiate death of the flea after a blood meal was taken. Similar types of vaccines have been tested in cattle infested with ticks, with fairly encouraging results for both ectoparasites.

COMPARATIVE MEDICINE CONSIDERATIONS

It is uncommon for humans to develop FAD. However, recurrent bites from ectoparasites in atopic individuals have the potential to induce IgE antibodies to salivary antigens. Abnormally severe and prolonged inflammation at the site of a bite from a flea, mosquito, or tick is indicative of a hypersensitivity response to salivary proteins from the ectoparasite.

FAD has been cited as the most common allergic disease in the cat. Cats that are infested with fleas in the absence of an immune response will show a small red mark at the site of the bite, whereas those that have mounted an IgE-mediated or delayed-type hypersensitivity response to the bite will develop a much more severe response. Flea-allergic cats show intense pruritus, like that observed in dogs, but in cats the lesions and their location are more variable. Four common reaction patterns are found in cats with allergic skin disease—miliary dermatitis, symmetrical alopecia, head and neck excoriations, and eosinophilic granuloma complex. These patterns are usually seen alone, but there have been cases of flea-allergic cats presenting with a combination of these patterns. Flea-allergic cats with miliary dermatitis have multiple small to large crusted papules that are located on the caudal dorsum, the thighs, or in a generalized pattern. In one study it was reported that 14% of cats with FAD presented with the single clinical sign of eosinophilic granuloma complex— a syndrome that is often called "rodent ulcer." Lesions around the neck and ears are also commonly seen. Cats with symmetrical alopecia exhibit hair loss bilaterally, and cats with head and neck excoriations have scabs and lesions behind the ears and over the ventral and dorsal neck area.

Intradermal testing for flea allergy in cats is commonly undertaken using whole body flea antigen (1/1000 weight/volume) injected intradermally into shaved skin, with saline and histamine controls. Cats should not be treated with corticosteroids for at least 4–8 weeks or with antihistamines for at least 10 days prior to the testing. The observation of the skin test site must include the 15-minute time point for the type I hypersensitivity response and a 48-hour observation to evaluate type IV reactivity. As with dogs, some cats show type I hypersensitivity, some show type IV, and others may respond at both time points, demonstrating both IgE- and cell-mediated reactivity to flea allergens. Serum testing is available, but cats that show only a type IV response will test negative on enzyme-linked immunosorbent assay (ELISA). Treatment of affected cats includes flea avoidance or flea control, as well as glucocorticoids to treat pruritus.

Horses do not become infested with fleas; however, hypersensitivity to biting insects is quite common. The bite of the ectoparasitic *Culicoides* gnats is commonly associated with development of a hypersensitivity dermatitis or "sweet itch" (see Case 16.) Other flies, such as black flies and stable flies, can induce type I hypersensitivity in genetically susceptible horses. In a similar way to dogs and cats with FAD, the affected horses develop IgE antibodies that are specific for the protein antigens present in fly saliva. They show severe pruritus, often rubbing out their mane and dorsal tail hair as a result of scratching. Measurement of IgE antibodies to fly allergens has confirmed that this condition is IgE mediated. There are also genetic factors that contribute to the development of sweet itch. The role of MHC class II molecules in antigen presentation is a well-recognized immunological function. Certain MHC class II specificities (ELA in horses) have been associated with the development of *Culicoides* hypersensitivity in Icelandic Horses and Exmoor Ponies.

Questions

1. Describe the process by which a dog develops hypersensitivity to flea antigens.

2. Compare the lesions and the immunological reactions that contribute to the formation of those lesions in dogs with a type I-mediated versus a type IV-mediated flea allergy dermatitis.

3. Explain the rationale for each of the two immune therapy vaccines that are mentioned in this case for flea allergy dermatitis.

Further Reading

Bruet V, Bourdeau PJ, Roussel A et al. (2012) Characterization of pruritus in canine atopic dermatitis, flea bite hypersensitivity and flea infestation and its role in diagnosis. *Vet Dermatol* 23:487–493.

Hobi S, Linek M, Marignac G et al. (2011) Clinical characteristics and causes of pruritus in cats: a multicenter study on feline hypersensitivity-associated dermatoses. *Vet Dermatol* 22:406–413.

Wilkerson MJ, Bagladi-Swanson M, Wheeler DW et al. (2004) The immunopathogenesis of flea allergy dermatitis in dogs, an experimental study. *Vet Immunol Immunopathol* 99:179–192.

Wuersch K, Brachelente C, Doherr M et al. (2006) Immune dysregulation in flea allergy dermatitis—a model for the immunopathogenesis of allergic dermatitis. *Vet Immunol Immunopathol* 110:311–323.

CASE 29
MULTIPLE MYELOMA

Plasma cells are the end stage of the B-lymphocyte lineage. When B lymphocytes are stimulated by antigen they undergo proliferation and differentiation, expanding the clone of B cells with receptors specific for the antigen that initiated the stimulation. In a normal immune response, a B cell stimulated by antigen binding to the B-cell receptor (BCR) also receives co-stimulatory signals from the binding of CD40 to CD40 ligand on the activated T-helper cell. A third stimulatory signal comes from the binding of the cytokine interleukin 4 (IL-4) to its receptor on the B cell. These signals trigger clonal expansion and production of memory B cells and antigen-producing plasma cells. The plasma cells make antibody that is specific for the same antigenic epitope as that which bound to the original B lymphocyte (Figure 29.1).

At any one time there are multiple B-cell clones that are activated to initiate the production of antibodies. Many chronic diseases—such as heartworm infestation, pyometra, and feline infectious peritonitis—cause so much B-cell activation that they induce many different clones to proliferate, making large amounts of antibody (polyclonal gammopathy). These antibodies are functional—in other words, they are directed towards epitopes on antigens that are associated with the infectious agent. In contrast, when a B-cell line undergoes malignant transformation, a plasma cell tumor results in the production of large amounts of monoclonal antibodies (called paraproteins), which are not directed against any specific pathogen that the patient has encountered. This large amount of antibody can be detrimental to the patient's health because it increases the viscosity of the blood and competes with legitimately stimulated B cells for bone-marrow space and resources to produce appropriate plasma cells and hematopoietic cells for erythrocyte and platelet production. Sometimes this malignant transformation occurs in bone-marrow plasma cells, and can even extend into peripheral lymphoid tissues. The malignant plasma cells crowd out normal B-cell development and decrease the levels of normal plasma cells. In addition to occupying space, these cells produce antibodies whose specificity is irrelevant to protection from pathogens, and whose presence in large amounts interferes with normal body function, causing clinical disease.

THE CASE OF MILLIE: A DOG FOR WHOM A ROUTINE BLOOD TEST REVEALED A SIGNIFICANT PROBLEM

SIGNALMENT/CASE HISTORY

Millie was a 12-year-old spayed female Australian Shepherd mix (Figure 29.2). She had been mostly well, apart from having had a rectal adenocarcinoma which had been successfully removed 5 years earlier. Her owner noticed a

TOPICS BEARING ON THIS CASE:

Plasma-cell malignancy

Paraproteins

Monoclonal gammopathy

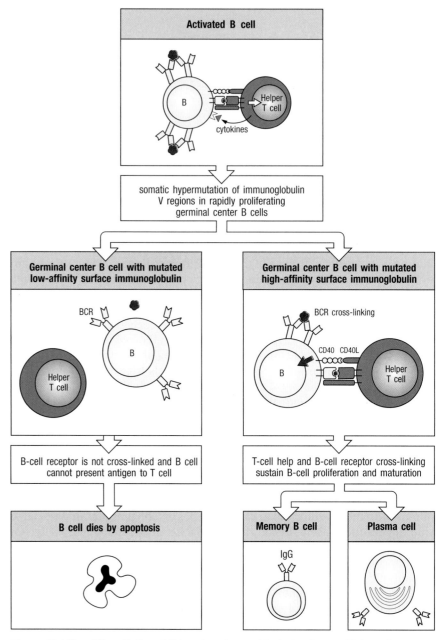

Figure 29.1 The differentiation of B lymphocytes and their development into mature plasma cells producing antibody. (From Murphy K [2011] Janeway's Immunobiology, 8th ed. Garland Science.)

Figure 29.2 Millie at 12 years of age.

swelling below Millie's right eye, and took her to the veterinarian. On examination it became apparent that the swelling was caused by a tooth abscess. An appointment was made for dentistry to treat the tooth abscess, and blood was taken as part of a routine pre-anesthetic work-up. The blood tests revealed abnormalities that were indicative of the presence of a gammopathy, so further investigations were undertaken.

PHYSICAL EXAMINATION

On physical examination, Millie was bright, alert, and responsive. Her mucous membranes were pale. There was a swelling in the area of the right upper premolar, and it was tender to the touch. Abdominal palpation revealed both hepatomegaly and splenomegaly.

DIFFERENTIAL DIAGNOSIS

Even before the results of the complete blood count (CBC) were obtained, it could be assumed that Millie was anemic because she had very pale mucous membranes. Anemia may be aplastic, immune mediated, or secondary to multiple myeloma. Since in Millie's case there is no history of acute blood loss, the erythrocytes either are not being produced, or are being destroyed. The enlarged liver and spleen would also be consistent with a diagnosis of multiple myeloma.

DIAGNOSTIC TESTS AND RESULTS

The results of the CBC supported the diagnosis of anemia. The total red blood cell count was $3.83 \times 10^6/\mu L$ (normal range, 5.6–$8.0 \times 10^6/\mu L$); the hemoglobin level and hematocrit were far below the normal range at 9.7 mg/dL (normal range, 14–19 mg/dL) and 28.2% (normal range, 40–55%), respectively. The presence of reticulocytes and nucleated red blood cells indicated that it was a regenerative anemia. The lymphocyte count was quite low at $663/\mu L$ (normal range, 1000–$4000/\mu L$). The total serum protein concentration was high at 15.3 g/dL (normal range, 5.4–7.4 g/dL), and the albumin/globulin ratio was low at 0.10 (normal range, 1–1.2), which was highly suggestive of multiple myeloma due to the very high levels of globulins. Urinalysis showed a +1 protein.

Additional tests were ordered based on the chemistry panel results and because of the unexpected abnormalities on the blood tests and physical examination. These tests—which would be expected to confirm or refute a diagnosis of multiple myeloma—included serum electrophoresis, immunoelectrophoresis, immunoglobulin quantitation, splenic aspirate, bone-marrow aspirate, and radiographic examination of all the long bones. The bone-scan results showed multiple lucent lesions in the intramedullary and endosteal regions of all the long bones (Figure 29.3A), and lytic lesions in the vertebral bodies (Figure 29.3B). These lesions are quite typical of multiple myeloma. The densitometry tracing showed a high narrow peak in the γ region (Figure 29.4). The bone-marrow aspirate contained increased numbers of plasma cells; both mature and immature cells were observed. Hematopoietic cells were normal in appearance (Figure 29.5), and the splenic aspirate contained large numbers of pleomorphic plasma cells. Immunoelectrophoresis showed a band which indicated that the immunoglobulin class being produced by the malignant plasma cells was IgG (Figure 29.6).

The single radial diffusion (SRD) test for quantitative immunoglobulins showed that the IgG concentration was 225 mg/mL, IgM was < 0.31 mg/mL, and IgA was < 0.10 mg/mL. It is typical of the immunoglobulin classes, which are not the

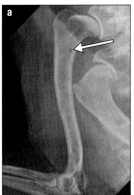 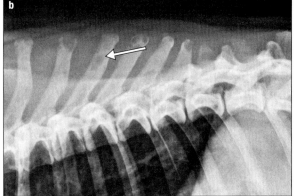

Figure 29.3 (a) Bone scan of Millie's humerus. (b) Bone scan of Millie's vertebrae. In both scans there are multiple lucent lesions (indicated by arrows). (Courtesy of Erik Wisner.)

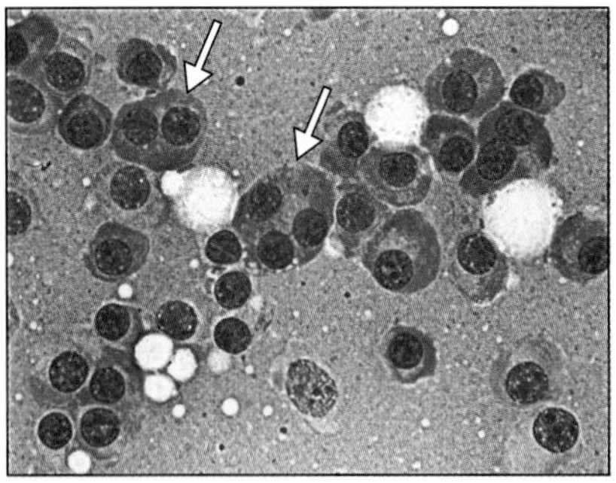

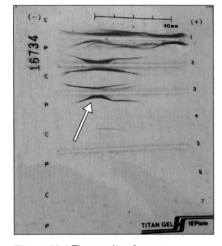

Figure 29.5 Numerous malignant plasma cells in bone-marrow aspirate. The arrows indicate dividing cells. (Courtesy of William Vernau.)

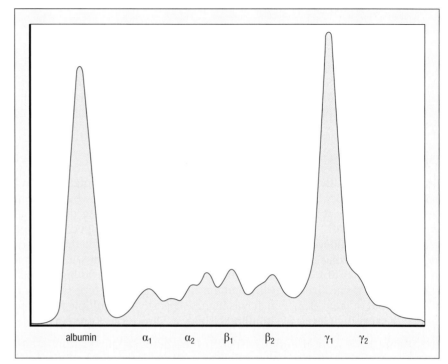

Figure 29.4 Densitometry tracing of serum electrophoresis showing a monoclonal spike in the γ region.

same class as the myeloma protein, to be reduced in concentration, whereas the class that is being produced by the malignant plasma cells is present at much higher concentrations than normal. A Bence Jones urinary protein test was negative. Bence Jones proteins are immunoglobulin light chains. They are sometimes, but not always, seen in plasma cell tumors and multiple myeloma.

DIAGNOSIS

Millie was diagnosed with multiple myeloma on the basis of the presence of malignant plasma cells in the bone marrow and spleen, lytic lesions in the long bones and vertebral bodies, and a monoclonal spike on serum densitometry, which was confirmed as IgG by single radial immunodiffusion (SRID) and immunoelectrophoresis.

TREATMENT

Millie was treated with melphalan, 2 mg (0.1 mg/kg) twice daily, and prednisolone, 10 mg once daily. She was re-examined weekly, and 3 weeks later her plasma protein concentration had decreased to 8.3 g/dL and her albumin:globulin ratio had risen to 0.43. She was still slightly anemic. After 1 month of treatment the splenomegaly was no longer palpable; the frequency of prednisolone was decreased to every other day. By week 7 after diagnosis, Millie was in remission (IgG concentration < 20 mg/mL). She was started on pulse therapy (treated for 1 week of each 4–6 weeks). She did well for several months and then presented with acute lameness with no history of trauma. A pathological fracture of the right tibia was diagnosed. Millie was euthanized because of the poor prognosis.

MULTIPLE MYELOMA

Multiple myeloma is a malignancy of plasma cells that occurs in multiple sites of lymphoid tissue in the body, but most commonly in bone marrow, spleen, liver, and lymph nodes. It is differentiated from a solitary tumor of plasma

Figure 29.6 The results of serum immunoelectrophoresis, which was performed to determine the isotype of the paraprotein. It identifies the monoclonal antibody (paraprotein) as IgG. Row C denotes the control serum well. Row P denotes Millie's serum well. The top reactions are between test sera and antibody to whole serum; the bottom reaction shows the dense compact arc formed by Millie's serum and antibody to the gamma chain of canine IgG (indicated by arrow).

cells (a plasmacytoma) by the fact that multiple organs are involved in multiple myeloma. Patients will usually present with high levels of plasma protein (consisting predominantly of paraprotein), anemia, and often renal complications due to the high viscosity of the plasma. Hypercalcemia is a frequent, although not constant, feature of the disease. Detection of high serum protein concentrations and the appearance of a monoclonal spike on the densitometry tracing from serum electrophoresis is usually the initial clue that multiple myeloma may be present. However, there are other causes of monoclonal antibodies. A monoclonal spike is sometimes associated with *Ehrlichia canis* and/or *Leishmania* infection. Quantitative immunoglobulin testing is appropriate, and generally identifies the antibody class of the paraprotein. In addition, the lack of normal levels of the remaining antibody classes is indicative of a predisposition to infection. Clinical presentation can include bone pain and even pathological fractures of long bones as the disease advances and bone lysis progresses. The presence of Bence Jones proteins (immunoglobulin light chains) in the urine is variable in canine cases. When present at high concentrations, these free light chains contribute to kidney pathology; proteinuria is common.

The antibody class of the paraprotein dictates the clinical presentation to some extent. Waldenström's macroglobulinemia, in which the paraprotein is IgM, produces the highest plasma viscosity, which may interfere with platelet function. Patients with this syndrome may present with bleeding abnormalities. IgA paraproteins also increase the viscosity of the blood, interfering with circulation, particularly in areas with small blood vessels.

COMPARATIVE MEDICINE CONSIDERATIONS

Multiple myeloma occurs in 1–4 per 100,000 people. Much like multiple myeloma in dogs, it causes bone pain and pathological fractures, anemia, and renal failure. Infections are also common in patients with multiple myeloma because even though there is a high concentration of gamma globulins, the antibodies are ineffective because they are from a single clone of plasma cells. As seen in Millie's quantitative immunoglobulin results, the antibody classes not affected by the malignancy are decreased. Melphalan is also used in the treatment of human myeloma patients, along with other therapies, including bone-marrow transplantation (hematopoietic stem cell transplantation). Remission is usually achieved after chemotherapy, but relapse is common.

Multiple myeloma has been reported in horses. The syndrome is very similar to that seen in human and canine cases. In cases reported in the literature, anemia, hyperproteinemia, and hypergammaglobulinemia were the most consistent findings on blood analysis. Clinical signs varied, but included anorexia, weight loss, bone pain, leg edema, and infection.

In cats there appear to be two different clinical presentations. One of these includes bone-marrow involvement and the radiographic presence of lytic lesions; the other lacks radiographic bone involvement, and includes the presence of extramedullary lesions, including skin tumors. In one study it was reported that 67% of cats with myeloma-related disease had extramedullary tumors, compared with only 5% of humans with the disease. In contrast, that study found that 8% of cats had radiographic lytic lesions in long bones or vertebrae, compared with 80% of human myeloma patients.

Questions

1. Explain the increased incidence of infection in a dog with an IgG level that is 10 times higher than normal.

2. What causes the high narrow spike on a densitometry tracing from a dog with multiple myeloma?

3. Why are patients with multiple myeloma usually anemic?

Further Reading

Edwards DF, Parker JW, Wilkinson JE & Helman RG (1993) Plasma cell myeloma in the horse: a case report and literature review. *J Vet Intern Med* 7:169–176.

Giraudel JM, Pagès JP & Guelfi JF (2002) Monoclonal gammopathies in the dog: a retrospective study of 18 cases (1986–1999) and literature review. *J Am Anim Hosp Assoc* 38:135–147.

Mellor PJ, Haugland S, Murphy S et al. (2006) Myeloma-related disorders in cats commonly present as extramedullary neoplasms in contrast to myeloma in human patients: 24 cases with clinical follow-up. *J Vet Intern Med* 20:1376–1383.

CASE 30
FELINE INFECTIOUS PERITONITIS

The immune response to viral infections usually involves both antibody production and induction of cell-mediated immune responses. Most often antibodies contribute toward decreasing viral propagation in the host and are helpful in ultimate control of the infection. In rare cases the antibody response to virus is not only ineffective in fighting viral infection, but also becomes destructive in its own right by promoting viral propagation and/or non-productive inflammation. One mechanism by which antibodies can be destructive is via excessive immune-complex formation, with complement fixation and deposition in blood vessels causing vasculitis. Another mechanism is called antibody-dependent enhancement, a process by which antibody binds to virus and then assists its entry into macrophages through Fc-receptor binding (Figure 30.1). In feline infectious peritonitis (FIP), for example, antibodies participate in disease enhancement by helping the coronavirus to gain entry to macrophages.

THE CASE OF DAFFY: A YOUNG CAT WHO DEVELOPED AN UNRESPONSIVE FEVER AND DIFFICULTY SEEING

SIGNALMENT/CASE HISTORY

Daffy was a 1-year-old Domestic Shorthair spayed female cat. She lived in a home with up to 10 other cats, as her owner provided a foster home for unwanted kittens. According to the owner, sometimes the kittens that she fostered had "crusty eyes and snotty noses." Daffy had been vaccinated for panleukopenia, rabies, leukemia, and calicivirus. She was presented to the veterinarian because she had begun to lose weight, was not eating well, and had become lethargic. Daffy's owner had previously taken her to another veterinarian, who had diagnosed a fever of 104°F and dispensed antibiotics that did not decrease the fever or resolve the other clinical signs. The owner also noted that the cat seemed to have difficulty seeing; her left eye had gradually become cloudy with a red discoloration (Figure 30.2).

PHYSICAL EXAMINATION

On examination, Daffy was responsive but lethargic, and had a dull unkempt hair coat. No external parasites were seen. Her temperature was 103.5°F (normal range, 100–102.5°F), her pulse was 196 beats/minute (normal range, 140–220 beats/minute), and her respiratory rate was 42 breaths/minute (normal range, 20–30 breaths/minute). Her nose was dry with a crusty discharge, and her pupils were dilated. There was no ocular discharge, and the corneas were clear. The

TOPICS BEARING ON THIS CASE:

Polyclonal gammopathy

Virus-induced secondary immunodeficiency

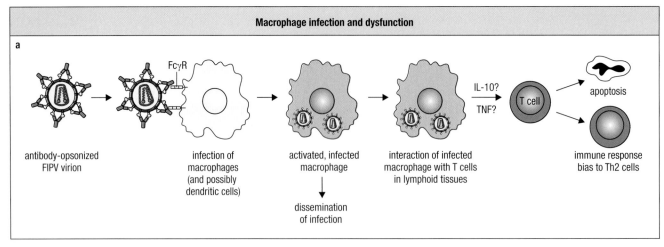

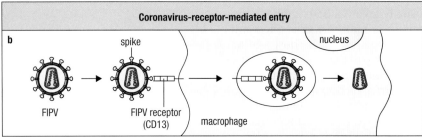

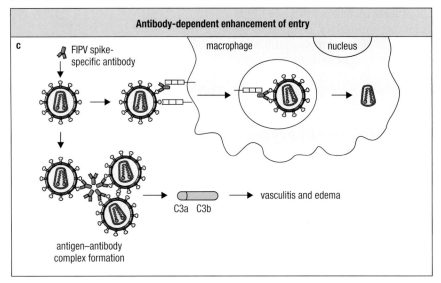

Figure 30.1 The immune enhancement of infection by the feline infectious peritonitis virus (FIPV). (a) Antibody to the spike protein can opsonize the virion for entry into the macrophage, where it can be disseminated throughout the body. (b) Normal entry is facilitated by binding of the spike protein to its receptor CD13. (c) Formation of antigen–antibody complexes can cause complement activation, with resultant immune-complex deposition and vasculitis. (Adapted from Dandekar AA & Perlman S [2005] *Nat Rev Immunol* 5:917–927.)

peripheral lymph nodes were not enlarged, and abdominal organs palpated within normal limits. On thoracic auscultation the lung fields were clear, but a grade II/VI systolic heart murmur was heard in the left paracostal area.

A full ophthalmological examination was performed to evaluate Daffy's eyesight. Her pupils were mydriatic; the menace reflex was absent, and there was mild scleral injection and keratic precipitates. The examination revealed a +1/4 aqueous flare, inflammatory infiltrates, and pre-retinal hemorrhage. It was concluded that the ophthalmic changes were almost certainly a manifestation of systemic disease.

DIFFERENTIAL DIAGNOSIS

The first conditions to be ruled out for the fever of unknown origin and the ocular changes included feline leukemia virus (FeLV), feline immunodeficiency virus (FIV), toxoplasmosis, and *Cryptococcus*. Feline infectious peritonitis (dry form) was high on the differential diagnosis list because of the ocular manifestations.

DIAGNOSTIC TESTS AND RESULTS

The results of a complete blood count (CBC) showed that Daffy had a mild regenerative anemia with a red blood count of $6.64 \times 10^6/\mu L$ (normal range, $7.0–10.5 \times 10^6/\mu L$), a hematocrit of 27.7% (normal range, 30–50%), and high plasma protein levels of 10 g/dL (normal range, 6.8–8.3 g/dL). The leukocyte count was high normal with a mild neutrophilia, including an increase in band neutrophils indicating increased neutrophil production. Abnormal parameters on the chemistry panel included slightly elevated glucose levels, a total protein concentration of 10.8 g/dL (normal range, 6.6–8.4 g/dL), a globulin concentration of 8.4 g/dL (normal range, 2.9–5.3 g/dL), a decreased albumin/globulin ratio of 0.31, a broad-based peak in the gamma-globulin region identified on the serum electropherogram, consistent with polyclonal gammopathy (Figure 30.3), and a bilirubin concentration of 0.6 mg/dL (normal range, 0–0.2 mg/dL).

The results of other tests indicated that the cat was FeLV negative, FIV negative, *Cryptococcus* negative, and toxoplasmosis negative. An abdominal ultrasound examination revealed the presence of an enlarged spleen. Ultrasound-guided biopsy identified splenitis: the spleen was infiltrated with mixed cell types, predominantly neutrophils. Mesenteric lymphadenopathy was also observed. These findings were suggestive of FIP. A serological test was performed to determine the antibody titer to feline coronavirus (FIP). It was positive at 1:6400, and a value of this magnitude was very suggestive of FIP. Lower titers can be found in cats that have been exposed to feline enteric coronavirus (FECV) and do not have FIP.

DIAGNOSIS

Daffy was diagnosed with feline infectious peritonitis (dry form). Diagnosis is difficult because an antibody test for FECV antibodies is positive in most cats and can be useful only if the titer is very high and the cat has the typical polyclonal gammopathy and the appropriate clinical signs. If the wet form of FIP is suspected, examination of effusion fluid for macrophages containing the coronavirus is the preferred way to confirm the diagnosis. Some laboratories offer a reverse-transcription polymerase chain reaction (RT-PCR) test for detection of the viral antigen in inflammatory fluids.

TREATMENT

Unfortunately there is no treatment for FIP. Daffy received supportive care until she was deemed to be too uncomfortable, and was then humanely euthanized.

FELINE INFECTIOUS PERITONITIS

FIP is a fatal disease of young cats caused by the feline enteric coronavirus (FECV), a double-stranded RNA virus. The virus primarily affects cats under 16 months of age, particularly those from catteries, shelters, or households with multiple cats. It is reported to be present in approximately 40% of kittens that come to animal shelters, and is often subclinical, causing at most a few days of loose stool or diarrhea. However, it can be shed for 4–6 months, sometimes intermittently, thus contaminating the environment and fostering transmission to other kittens. FIP is caused by a mutant form of FECV; this mutation is thought to occur within the FECV-infected kitten. Only a small percentage

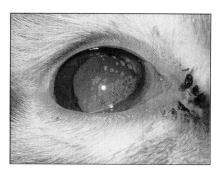

Figure 30.2 Daffy displaying ocular manifestations of the dry (non-effusive) form of FIP. (From Schaer M [2009] Clinical Medicine of the Dog and Cat, 2nd ed. Courtesy of CRC Press.)

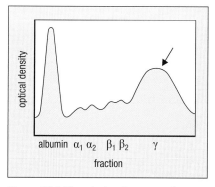

Figure 30.3 The electropherogram from serum electrophoresis showing the typical polyclonal gammopathy seen with chronic inflammation, as in FIP infection. The γ peak (indicated by arrow) is higher and broader than normal. In addition, the albumin peak in this FIP-infected cat is smaller than normal. These altered serum protein parameters are responsible for the decreased albumin/globulin ratio.

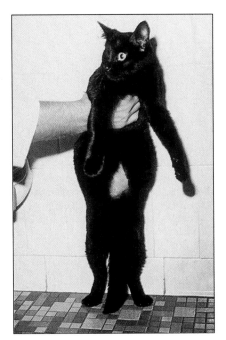

Figure 30.4 A young cat with a peritoneal effusion that is typical of the "wet" or effusive form of FIP. (From Schaer M [2009] Clinical Medicine of the Dog and Cat, 2nd ed. Courtesy of CRC Press.)

of cats that harbor a mutant FIP virus actually develop disease, most commonly when the mutation occurs in a gene called *3c*. The FIP disease itself is not highly contagious, as each affected cat appears to harbor its own unique mutated virus in its body tissues. According to some studies, genetics play a major role in FIP, with both maternal and paternal susceptibility factors contributing to the heritability of the disease. Stress is also thought to be a factor. All breeds of cats can be affected, but some veterinarians claim that there is an increased incidence in Burmese cats.

Most cats that develop FIP are less than 4 years of age. The clinical signs occur within weeks to months after the initial infection. As cats become affected they show nonspecific signs of sickness, which can include anorexia, poor hair coat, weight loss, lethargy, and secondary infections. Often there is a fever that is not responsive to antibiotic therapy. There are two forms of the disease—effusive or "wet," and non-effusive/granulomatous or "dry." The wet form is more common, and most veterinarians easily recognize the signs in a wasted kitten with a huge belly containing ascites fluid (Figure 30.4), or with dyspnea due to fluid accumulation in the pleural cavity. The dry form of the disease is less easily recognized as it is not associated with fluid accumulations, but instead with masses in the liver, spleen, kidney, or other organs, including the nervous system and the eye. In both forms of the disease, blood parameters show anemia in over 50% of cases, and a left shift in neutrophils is common. The blood chemistry usually shows hypoalbuminemia and hyperglobulinemia, and icterus and uveitis are also common. As the disease progresses, the cat becomes severely immunocompromised and may develop secondary infections and/or neurological symptoms.

Diagnosis of FIP on the basis of laboratory test results relies on detection of anemia on the CBC, elevated serum protein levels, and the presence of a polyclonal gammopathy. The immune response to FIP virus is an exuberant humoral response in which the antibodies produced are not protective. Some experimental studies have shown that they can also bind to virus and facilitate its entry into macrophages. The macrophages fail to kill the virus, and instead transport it throughout the body and become a source of inflammation. This "antibody-dependent enhancement" response has been demonstrated in experimental infections, but has not been confirmed to be a major pathogenic mechanism in naturally occurring cases of FIP. There is a vaccine on the market for FIP, but its use is not currently recommended because of the likelihood that non-virus-neutralizing antibodies will be induced and cause antibody-dependent enhancement. A recent study has shown that in cats naturally infected with FIP, both natural killer (NK) cells and T-regulatory cells are depleted from the blood, mesenteric lymph node, and spleen. This means that cats with FIP have severe immunosuppression of the important NK innate cells that would normally assist in antiviral defense. The depression of T-regulatory cells is also probably a mechanism that helps to facilitate the immunopathological response that occurs in FIP. The granulomatous response that occurs with FIP infection is characterized by neutrophil infiltration, which is not usually a hallmark of viral infection. In an *in-vitro* study of feline macrophages infected with FIP virus and specimens from FIP-infected cats, macrophages targeted by FIP virus produced tumor necrosis factor-alpha (TNFα), granulocyte-macrophage colony-stimulating factor (GM-CSF), and granulocyte colony-stimulating factor (G-CSF), all of which are neutrophil survival factors. Thus the infected macrophages may set the conditions that influence the development of the inflammatory lesions.

COMPARATIVE MEDICINE CONSIDERATIONS

FIP is an example of a disease that causes a polyclonal gammopathy. This antibody response is not only ineffective in eliminating the virus, but also very probably contributes to the persistence of the virus and its dissemination throughout the body. Humans can be infected by a virus called dengue, and in the resultant disease the immune response shows a similar pattern to that of FIP. In dengue virus infection, a patient who has previously made antibodies to a different serotype of dengue virus is infected with a second serotype. The virus is not neutralized, but instead taken into the macrophages by Fc-receptor binding, facilitating viral replication and spread. This disease is probably the most well-studied example of antibody-mediated disease enhancement in humans.

Questions

1. A high percentage of cats are infected with the enteric coronavirus. Why do only a few of them develop FIP?

2. Why is FIP so difficult to diagnose using serology? What would be a better test?

3. What are the three primary effects of FIP on the immune system?

Further Reading

Riemer F, Kuehner KA, Ritz S et al. (2016) Clinical and laboratory features of cats with feline infectious peritonitis – a retrospective study of 231 confirmed cases (2000–2010). *J Feline Med Surg* 18:348–356.

Takano T, Azuma N, Satoh M et al. (2009) Neutrophil survival factors (TNF-alpha, GM-CSF, and G-CSF) produced by macrophages in cats infected with feline infectious peritonitis virus contribute to the pathogenesis of granulomatous lesions. *Arch Virol* 154:775–781.

Tekes G & Thiel HJ (2016) Feline coronaviruses: pathogenesis of feline infectious peritonitis. *Adv Virus Res* 96:193–218.

Vermeulen BL, Devriendt B, Olyslaegers DA et al. (2013) Suppression of NK cells and regulatory T lymphocytes in cats naturally infected with feline infectious peritonitis virus. *Vet Microbiol* 164:46–59.

CASE 31
MONOCYTIC
EHRLICHIOSIS

Serum electrophoresis is performed as part of a clinical chemistry analysis. The technique involves exposing serum proteins to an electric current, which separates the proteins on the basis of their charge. The resulting bands are then translated using a densitometer, and the result is visualized as an electropherogram (Figure 31.1A). Serum proteins separate into α_1 and α_2, β_1 and β_2, and γ fractions. In some diseases there is an abnormal distribution of the serum proteins into these fractions. Gammopathy is defined as an abnormality in the gamma-globulin fraction of the serum, whereas polyclonal gammopathy refers to an elevated gamma-globulin fraction that appears as a broad band on the densitometry tracing (Figure 31.1B). The latter occurs when many different plasma cells produce antibodies that differ in their heavy and light chain variable regions because they result from the activation of many different B cells. There are many possible causes of polyclonal gammopathy, including chronic infection, some autoimmune diseases, and neoplasia. A monoclonal gammopathy is most often associated with a malignancy of plasma cells, which display a clonality that results in production of high levels of identical antibody molecules. In this case the antibodies are formed as a result of the proliferation of a single clone of B lymphocytes with similar heavy and light chain variable and constant regions (Figure 31.1C). Monoclonal gammopathy is commonly associated with a plasma cell tumor or multiple myeloma.

In some of the tick-borne diseases, such as ehrlichiosis, both polyclonal and mixed gammopathy (restricted oligoclonal gammopathy) have been observed in infected dogs. Canine monocytic ehrlichiosis in particular is known to cause both polyclonal and monoclonal gammopathy. The disease is the result of an infection with the obligate intracellular rickettsia *Ehrlichia canis*, transmitted via the bite of the tick *Rhipicephalus sanguineus*. It exists in a subacute (asymptomatic), acute, and chronic form, and is difficult to diagnose because it is associated with multiple clinical manifestations. Because *E. canis* is an obligate intracellular parasite, the antibody response is not an effective defense. For this type of disease agent, a T-helper type 1 response is required. Activated $CD4^+$ T cells producing interferon γ will activate monocytes with intracellular *E. canis* to become effective killers.

THE CASE OF JORDAN: A DOG WITH BLOOD-TINGED URINE, FEVER, AND DEPRESSION

SIGNALMENT/CASE HISTORY

Jordan is a 5-year-old castrated male yellow Labrador Retriever (Figure 31.2) who lives in a wooded area in the northeastern United States. He was presented to his veterinarian one summer day with acute onset of lethargy, anorexia, high

TOPICS BEARING ON THIS CASE:

Serum protein electrophoresis; gammopathy—polyclonal and restricted oligoclonal

Inflammation

Immune response to an obligate intracellular bacterium

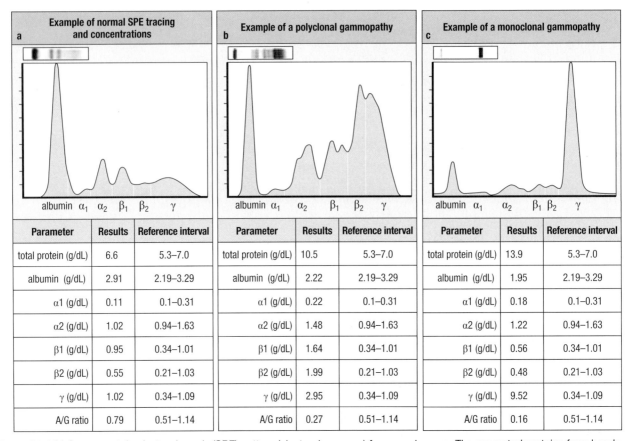

a	Example of normal SPE tracing and concentrations		
Parameter	Results	Reference interval	
total protein (g/dL)	6.6	5.3–7.0	
albumin (g/dL)	2.91	2.19–3.29	
α1 (g/dL)	0.11	0.1–0.31	
α2 (g/dL)	1.02	0.94–1.63	
β1 (g/dL)	0.95	0.34–1.01	
β2 (g/dL)	0.55	0.21–1.03	
γ (g/dL)	1.02	0.34–1.09	
A/G ratio	0.79	0.51–1.14	

b	Example of a polyclonal gammopathy		
Parameter	Results	Reference interval	
total protein (g/dL)	10.5	5.3–7.0	
albumin (g/dL)	2.22	2.19–3.29	
α1 (g/dL)	0.22	0.1–0.31	
α2 (g/dL)	1.48	0.94–1.63	
β1 (g/dL)	1.64	0.34–1.01	
β2 (g/dL)	1.99	0.21–1.03	
γ (g/dL)	2.95	0.34–1.09	
A/G ratio	0.27	0.51–1.14	

c	Example of a monoclonal gammopathy		
Parameter	Results	Reference interval	
total protein (g/dL)	13.9	5.3–7.0	
albumin (g/dL)	1.95	2.19–3.29	
α1 (g/dL)	0.18	0.1–0.31	
α2 (g/dL)	1.22	0.94–1.63	
β1 (g/dL)	0.56	0.34–1.01	
β2 (g/dL)	0.48	0.21–1.03	
γ (g/dL)	9.52	0.34–1.09	
A/G ratio	0.16	0.51–1.14	

Figure 31.1 (a) Serum protein electrophoresis (SPE) pattern (electropherogram) for normal serum. The separated proteins form bands after separation. These are shown above (in blue). Albumin (on the far left) is the densest band because it is present at the highest concentration in the serum. The albumin/globulin (A/G) ratio is calculated, and is normally between 0.5 and slightly more than 1.0. (b) Electropherogram of a polyclonal gammopathy. Note that the β$_2$ and γ fractions are much larger than normal. The A/G ratio at 0.27 is very low. This type of polyclonal gammopathy is seen in autoimmune and chronic infectious disease. (c) Electropherogram of serum from a patient with multiple myeloma, showing a very large γ-globulin spike. The fraction is narrow because all of the immunoglobulin that is being produced results from a single clone of plasma cells, and thus the proteins are identical. Note the very low albumin concentration and very low A/G ratio (0.16). (Adapted courtesy of ECLINPATH, Cornell University.)

fever, and depression. The owner reported that Jordan's urine had a blood tinge daily (hematuria), but not with every urination. There was no report of known toxin exposure, and no tarry or bloody stools had been observed. The owner commented that he frequently plucked ticks off Jordan.

PHYSICAL EXAMINATION

On physical examination, Jordan was depressed, with a temperature of 103.5°F (normal range, 100.5–102°F). His body condition was good, with adequate hydration. No abnormalities were noted in any of the organ systems examined. The prostate was of a normal size, and no calculi were palpated in the urethra. The oral mucous membranes were slightly pale, and the capillary refill time was normal at 2 seconds. Abdominal palpation did not elicit pain. The liver was palpated just inside the costal arch, and the spleen was prominent.

DIFFERENTIAL DIAGNOSIS

Hematuria has a number of potential causes, including urinary tract infection, urinary tract neoplasia, urolithiasis, and prostate disease. Other less common causes include trauma, clotting disorders, and idiopathic benign

Figure 31.2 Jordan in his backyard before he became ill. (Courtesy of Shutterstock, copyright Alexx60.)

renal hematuria. It is important to determine in this case whether the blood observed by the owner is frank blood, or whether the pink tinge is caused by hemoglobinuria resulting from hemolysis.

DIAGNOSTIC TESTS AND RESULTS

A complete blood count (CBC), blood chemistry, and urinalysis were performed to establish a basis for further testing. The hematocrit was 25% (normal range, 40–50%), with a mean cell volume (MCV) of 65 and a mean corpuscular hemoglobin concentration (MCHC) of 33 g/dL. Reticulocytes were increased. The white blood cell count of 8400 cells/μL was within normal limits, but the platelet count was low at 90,000/μL (normal range, 150-400/μL). These results indicate a mild anemia and thrombocytopenia. The results of blood chemistry showed a decreased albumin/globulin (A/G) ratio of 0.12. An increase in the amount of globulin (12.9 g/dL, compared with a value of 1.5 g/dL for albumin) was responsible for the low A/G ratio, and indicated the presence of a gammopathy. Gammopathy was observed on the electropherogram, and appeared as a restricted oligoclonal type (Figure 31.3). All other parameters were within normal limits. The urinalysis showed that 4+ occult blood was present in the urine sample (which had been obtained by free catch). There were more than 50 red blood cells per high-power field, the specific gravity was 1.015, and there was 3+ protein in the urine (normally neither blood nor protein are present in urine).

On the basis of these results, additional tests were performed, including a urine culture, which was negative, and an abdominal ultrasound examination. There was no evidence of urethral calculi or bladder stones, but an enlarged liver and spleen were noted. Ultrasound-guided aspiration was performed on both the liver and the spleen. The splenic aspirate showed a mature granular lymphoproliferative disease with plasma cell hyperplasia. The liver aspirate showed a marked mature lymphoproliferative disease. In view of the dog's history of exposure to ticks and the noted pathology, a test was performed for exposure to tick-borne organisms, namely the SNAP® 4Dx® four-way test for tick-borne disease. The test was positive for antibodies to *Ehrlichia canis*. A serum sample was then submitted to the laboratory to determine a titer to *E. canis*. The titer result was very high (1:20,480), indicating the presence of ongoing infection.

Causes of polyclonal gammopathy that need to be ruled out include immune-mediated disease such as systemic lupus erythematosus (SLE). Lymphoproliferative changes in the liver and spleen could have been attributed to the chronic *E. canis* infection, but these changes are also consistent with malignancy. T-cell lymphoma of the liver and spleen must therefore be considered. The definitive test for *E. canis* infection is quantitative reverse-transcription polymerase chain reaction (RT-PCR), which in Jordan's case revealed infection.

DIAGNOSIS

Jordan was diagnosed with an *Ehrlichia canis* infection in the subacute phase. Diagnosis can sometimes be made by visualizing morula within white blood cells on a blood smear (Figure 31.4), but antibody assays, indirect immunofluorescence, and enzyme-linked immunosorbent assay (ELISA) are also commonly used to confirm infection. Some laboratories offer PCR diagnostics for *Ehrlichia* infection. The chronicity of the infection and the presence of the organism in the liver and spleen induce hypergammaglobulinemia, which is the result of a polyclonal plasma cell response.

TREATMENT

Jordan was treated with doxycycline (150 mg by mouth every 12 hours for 14 days), with eventual resolution of his disease. Other tetracycline antibiotics can also be successfully used as treatment. There is no vaccine for canine

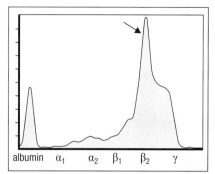

albumin α₁ α₂ β₁ β₂ γ

Figure 31.3 Electropherogram showing a monoclonal spike (indicated by arrow) superimposed on a polyclonal-like base. This is termed a "restricted oligoclonal pattern," and is frequently seen in *Ehrlichia canis* infection. Compare this with the polyclonal pattern that is shown for FIP infection in Figure 30.3, and the monoclonal pattern shown for multiple myeloma in Figure 29.4. (Adapted courtesy of ECLINPATH, Cornell University.)

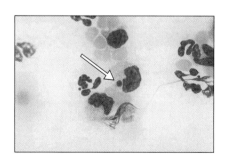

Figure 31.4 *Ehrlichia canis* can sometimes be diagnosed on a blood smear when monocytes containing morula (indicated by arrow) are present. (From Shaw S & Day M [2005] Arthropod-borne Infectious Diseases of the Dog and Cat. Courtesy of CRC Press.)

ehrlichiosis, and reinfection can occur. Prevention of further tick bites is therefore important.

MONOCYTIC EHRLICHIOSIS

Canine monocytic ehrlichiosis is caused by infection with *Ehrlichia canis*, a small non-motile obligate intracellular organism that infects monocytes and is transmitted by the brown dog tick (*Rhipicephalus sanguineus*). Less commonly it can be caused by *E. chaffeensis* and *E. ewingii*, both of which have zoonotic potential. A subclinical phase can occur prior to acute disease. The initial acute disease usually occurs within 1–3 weeks after infection, is characterized by fever, anorexia, lethargy, and often lymphadenopathy, and may progress to thrombocytopenia. If not cleared, this stage is followed by a subacute phase associated with hypergammaglobulinemia (usually polyclonal or restricted oligoclonal but sometimes monoclonal gammopathy), anemia, and thrombocytopenia. The subacute form can last for many months. *E. canis* infection can sometimes present as acute blindness. In these cases there is often a high plasma viscosity associated with a monoclonal gammopathy; platelet dysfunction and thrombocytopenia are implicated in ocular bleeding. The hematuria observed on presentation in this case is similarly the result of severe thrombocytopenia, which can cause bleeding into a variety of organs, including the urinary bladder.

Without treatment, dogs infected with *E. canis* can progress to the chronic form of the disease. Clinical signs can include pancytopenia, bone-marrow suppression, and hemorrhage. Compared with other breeds, the German Shepherd has been associated with a higher incidence of progression to the chronic phase and consequent mortality when infected with *E. canis*. During the Vietnam War, many American military German Shepherd dogs died from the disease in Southeast Asia.

COMPARATIVE MEDICINE CONSIDERATIONS

Human monocytic ehrlichiosis is usually caused by *E. chaffeensis* transmitted by the lone star tick (*Amblyomma americanum*), although *E. canis* has been isolated from some human cases. White-tailed deer and domestic dogs serve as reservoirs. The symptoms of acute infection in humans include a moderate to high fever, lower back pain, gastrointestinal discomfort with or without vomiting, and diarrhea. Lymphadenopathy may also occur. Diagnosis involves similar techniques to those used in the canine disease, with RT-PCR being the preferred method.

Equine granulocytic ehrlichiosis is caused by infection with the rickettsia *Ehrlichia equi*, which has been renamed *Anaplasma phagocytophilum*, based on DNA sequencing (it became apparent that the organism is closer phylogenetically to *Anaplasma* species than to organisms in the genus *Ehrlichia*). The vector for this bacterium is the western black-legged tick (*Ixodes pacificus*), and infection appears to be primarily seasonal in the foothills of northern California, although seroprevalence has shown that exposure also occurs elsewhere in the United States. Humans can also develop granulocytic ehrlichiosis. However, neither this nor the equine form of the infection are very similar immunologically to canine and human monocytic ehrlichiosis.

Questions

1. What is the significance in this case of a very low albumin/globulin ratio?

2. *Ehrlichia canis* is an obligate intracellular organism that infects monocytes. What immune mechanism do you think would be most effective in controlling this infection? Why?

3. Systemic lupus erythematosus (Case 23), feline infectious peritonitis (Case 30), and canine monocytic ehrlichiosis (this case) are all often associated with a polyclonal gammopathy. Explain how and why these very different diseases induce a similar pattern of serum proteins.

Further Reading

Breitschwerdt EB, Woody BJ, Zerbe CA et al. (1987) Monoclonal gammopathy associated with naturally occurring canine ehrlichiosis. *J Vet Intern Med* 1:2–9.

Harrus S & Waner T (2011) Diagnosis of canine monocytotropic ehrlichiosis (*Ehrlichia canis*): an overview. *Vet J* 187:292–296.

Harrus S, Waner T, Avidar Y et al. (1996) Serum protein alterations in canine ehrlichiosis. *Vet Parasitol* 66:241–249.

Harrus S, Ofri R, Aizenberg I & Waner T (1998) Acute blindness associated with monoclonal gammopathy induced by *Ehrlichia canis* infection. *Vet Parasitol* 78:155–160.

Singh AK & Thirumalapura NR (2014) Early induction of interleukin-10 limits antigen-specific CD4[+] T cell expansion, function, and secondary recall responses during persistent phagosomal infection. *Infect Immun* 82:4092–4103.

Tajima T & Wada M (2013) Inhibitory effect of interferon gamma on frequency of *Ehrlichia canis*-infected cells *in vitro*. *Vet Immunol Immunopathol* 156:200–204.

APPENDIX I
VACCINES AND
VACCINATION
SCHEDULES

TYPES OF VACCINES

KILLED OR INACTIVATED VACCINES

Bacteria and viruses can be inactivated chemically or with heat so that their antigenicity remains intact but they are no longer able to cause disease. Vaccines composed of this inactivated material are generally incorporated into an adjuvant—that is, a substance that creates a depot of antigen in the tissues in order to prolong immune system exposure, and that may also stimulate a more robust response. The immune response induced by non-living vaccines consists primarily of antibody, which is the body's primary line of defense against bacterial pathogens and some viruses. The non-living antigens are taken into phagocytic cells and processed by the exocytic pathway for presentation to CD4+ T-helper cells on major histocompatibility complex (MHC) class II molecules. A common example of a killed viral vaccine is the rabies vaccine used in multiple species. A common example of a bacterin (an inactivated bacterial vaccine) is the *Histophilus somni* bacterin used in cattle.

Toxoids are inactivated toxins. They are used as vaccines when the disease that the animal is being vaccinated against is caused by the production of a toxin, and toxin-neutralizing antibodies are therefore needed to protect the animal. Like killed or inactivated bacterial and viral vaccines, toxoids are usually accompanied by an adjuvant. An example is the tetanus toxoid that is used to prevent tetanus in horses.

MODIFIED LIVE OR ATTENUATED VACCINES

Viral agents are intracellular pathogens, and therefore require some T-cell-mediated immunity (CD8+ cytotoxic T cells) in addition to antibody to rid the animal of infection. Some viruses are handled well by antibody alone, but those that are highly cell associated, such as herpesviruses, are able to avoid interaction with antibody. To combat these viruses, vaccines that are able to replicate in the animal without causing disease are the best choice. Antigen presentation with class I molecules is the key to induction of cell-mediated immunity by the endocytic pathway taken by these living agents that replicate in the cytosol. Many of the early vaccines for canine distemper were made less virulent by passaging them in alternative species; the weakened virus then became a vaccine. Attenuation in the modern era is alternatively accomplished by deleting one or more virulence genes and then using the modified virus in a vaccine. There is a vaccine for porcine pseudorabies that utilizes this method. It is not usually necessary to incorporate a modified live vaccine into an adjuvant.

SUBUNIT VACCINES

It can often be determined which part of a bacterium or virion contains the protective antigenic epitope. When one or more protective antigens are either purified from the pathogen or created by molecular cloning technology, they can be used as a subunit vaccine. There are many advantages to this approach:

- The vaccine cannot revert to virulence or be accidentally improperly activated because there is no whole pathogen present.

- There is no need to administer other parts of the pathogen that may induce antibodies that are not protective and might contribute to a pro-inflammatory response (for example, fever, malaise).

- It makes possible the development of a "DIVA" (Differentiating Infected from Vaccinated Animals) vaccine. The advantage of this type of vaccine is that it allows differentiation between antibodies made in response to vaccination and those made as part of an immune response to infection.

Subunit vaccines often (but not always) require an adjuvant. The outer surface protein A (OspA) vaccine for *Borrelia burgdorferi* (Lyme disease) is an example of a subunit vaccine.

VIRAL-VECTORED VACCINES

One of the newer approaches to viral vaccines combines the DIVA advantage of the subunit vaccine with the ability to induce a cell-mediated response inherent in the live attenuated vaccines. This technique usually involves a large vector virus (vaccinia virus, canarypox virus), although other viruses are being studied as potential vectors. First, these vector viruses have some non-essential genes removed. As with the subunit vaccines it is necessary to know which epitopes on the viral pathogen elicit protective immunity. Once this is known, the genes for those proteins are cloned and inserted into the genome of the vector. The entire vector virus is then injected into the patient as vaccine, usually without an adjuvant. The vector undergoes limited replication, during which time the genes for the cloned proteins are translated into protein, and those proteins are expressed on host cells. Examples of viral-vectored vaccines include canarypox-vectored distemper vaccine, vaccinia-vectored rinderpest vaccine, and vaccinia-vectored rabies vaccine (used in wildlife). These vaccines elicit both antibody and cell-mediated immune responses.

DNA VACCINES

A DNA vaccine is similar in principle to a vectored vaccine in that the genes encoding those epitopes against which an immune response is desired are inserted into DNA. However, in a DNA vaccine the receiving DNA is not incorporated into a virus/vector, but is simply a naked piece of DNA in a circular form (a plasmid). The DNA vaccine is usually injected into the muscle with an air-powered injector rather than a traditional needle; sometimes the DNA is coated onto small gold beads. DNA vaccines, like their vectored counterparts, induce antibody and cell-mediated immunity, although the cell-mediated immunity in some examples has predominated. There is only one example of a licensed DNA vaccine, namely the canine melanoma vaccine (Oncept®, Merial®). This contains a gene for human tyrosine kinase, an enzyme that is overexpressed in melanoma cells. By virtue of cross-reacting epitopes, the expression of the human gene stimulates a cell-mediated response against the vaccinated animal's own tumor cells. The vaccine is used to treat melanoma in dogs, and off label to treat melanoma in horses.

ROUTES OF VACCINATION

The most common route of vaccination for mammals is parenteral. The subcutaneous route is used in many species because it is less painful than vaccination by the intramuscular route and it does not cause tissue damage. Several vaccines intended to protect mucosal surfaces have been manufactured in a modified live virus preparation for intranasal or oral vaccination. These vaccines induce production of secretory IgA in the secretions. Administration of vaccine in the water supply is the most common way to vaccinate poultry and fish. In some cases there are vaccinations available for administration by multiple possible routes for protection against a condition. For example, there are parenteral, intranasal, and oral vaccines available to protect against canine kennel cough. Similarly, there are both parenteral and intranasal vaccines available to protect against equine influenza.

SMALL ANIMAL (DOG AND CAT) VACCINES AND VACCINATION SCHEDULES

In the past 10 years there has been a coordinated effort by several veterinary organizations to develop guidelines for the immunization of dogs and cats. These guidelines seek to define which vaccines should be routinely administered (core vaccines), which should be selected depending on the lifestyle of the animal (non-core vaccines), and what the appropriate frequency of re-immunization should be.

PRINCIPLES OF IMMUNIZATION FOR PUPPIES AND KITTENS

Dogs and cats obtain most of their maternal antibody through colostrum, and about 10% through the placenta. The IgG that is transferred provides protection to the neonate during the first few months of life. If the dam's core vaccinations were up to date, maternal antibodies are often still present at 12 weeks of life. However, if the dam was unvaccinated against core diseases, the puppy or kitten could be susceptible as early as a few weeks of age. Without measuring antibody titers in the dam, the usual approach to vaccination of the neonate with core vaccines (with the exception of rabies) is to begin vaccination by 6–8 weeks of age and to provide additional vaccinations at 3- to 4-week intervals until the young animal reaches the age of 16 weeks. Vaccination of a puppy or kitten that has a high maternal titer will not result in stimulation of an active immune response, and thus the vaccine is essentially cancelled by the pre-existing antibodies. Once the 16-week threshold has been reached, the next administration of those antigens will not occur until 1 year of age, and thereafter every 3 years. The rabies vaccine is given once at 4 months of age, and is repeated 1 year later. Thereafter rabies vaccination can be given every 3 years, unless otherwise legally specified in the animal's home state or country.

CORE AND NON-CORE VACCINES FOR DOGS AND CATS

The core vaccines for dogs, according to both the American Animal Hospital Association and the American Veterinary Medical Association, are listed in Table A.1. These are vaccines that all dogs should receive regardless of their lifestyle.

Non-core vaccines for dogs are also listed in Table A.1. These are vaccines that are not required for all dogs, and that are administered based on potential exposure to the pathogen against which the vaccine is intended to give protection. For example, a dog that regularly goes into areas that are infested with the tick vector of *Borrelia burgdorferi* should be given the Lyme disease vaccine,

Table A.1 Core and non-core vaccines for dogs

Core vaccines	Non-core vaccines
Canine distemper virus	Canine parainfluenza virus
Canine parvovirus type 2	*Bordetella bronchiseptica*
Canine adenovirus type 2	*Borrelia burgdorferi* (Lyme disease)
Rabies virus	*Leptospira interrogans* (combined with other species of *Leptospira*)

Table A.2 Core and non-core vaccines for cats

Core vaccines	Non-core vaccines
Feline parvovirus vaccine (FPV)	Feline leukemia virus
	Chlamydophila felis
Feline herpesvirus type 1 (FHV-1)	Feline immunodeficiency virus (FIV)
Calicivirus	*Bordetella bronchiseptica*
Rabies virus (if endemic or required)	

whereas dogs that are unlikely to be exposed to that vector do not require the vaccine. Dogs that are boarded in kennels and/or that frequent areas with large numbers of dogs (such as dog parks) should be given a kennel cough vaccine (*Bordetella bronchiseptica*/parainfluenza virus).

The guidelines also include a "do not give" list of vaccines that are either not effective or not worth giving because the prevalence and severity of the disease do not justify administration of a vaccine.[1] One example is the coronavirus vaccine.

The American Association of Feline Practitioners (AAFP) has developed a set of guidelines (last updated in 2013) for vaccination of cats. Core and non-core vaccines are listed in Table A.2.[2]

DURATION OF IMMUNITY: HOW OFTEN TO RE-IMMUNIZE

Recent recommendations on the frequency of vaccination have been based on studies which demonstrate that duration of immunity (DOI) for core vaccines is longer than 1 year. The DOI is based on the presence of memory T and B cells and long-lived plasma cells in the vaccinated animal. Current recommendations for canine and feline core vaccines are for 3 years after the initial 1-year boost that follows the puppy or kitten vaccines. Rabies vaccination will vary depending on the state laws (in the USA), but unless otherwise required a 3-year interval is suggested for rabies vaccination of dogs and cats.

EQUINE VACCINES AND VACCINATION SCHEDULES

Core and non-core vaccine recommendations for horses are listed in Table A.3. Non-core vaccines are selected for use in horses by risk assessment. Criteria for frequency of vaccination and appropriate use have been established by the American Association of Equine Practitioners.[3]

The primary equine vaccines include several vaccine types and administration methods. For example, for West Nile virus there is an inactivated and recombinant canarypox-vectored vaccine. For equine influenza there is a modified live virus vaccine that is administered by the intranasal route, inactivated vaccine, and a canarypox-vectored recombinant vaccine.

BOVINE VACCINES AND VACCINATION SCHEDULES

The leading cause of morbidity and mortality of dairy calves and feedlot steers is bovine respiratory disease (BRD). It is a complex disease, caused by an interaction of stress with bacterial and viral pathogens resulting in severe pneumonia. The primary disease agents for BRD (see also Case 13) include bovine respiratory syncytial virus (BRSV), bovine herpesvirus type 1, parainfluenza type 3 virus (PI3V), and bovine viral diarrhea virus (BVDV) accompanied by one or more bacterial pathogens (*Pasteurella multocida*, *Mannheimia haemolytica*, *Histophilus somni*, and *Mycoplasma bovis*). Stressors can include castration, shipping, and weaning. A variety of vaccines are marketed for BRD, and

Table A.3 Core and non-core vaccines for horses

Core vaccines	Non-core vaccines
Tetanus toxoid	Strangles (*Streptococcus equi*)
Eastern and western encephalitis	Potomac horse fever (*Rickettsia ristici*)
West Nile virus	Equine influenza
Rabies	Equine herpesvirus
	Equine viral arteritis
	Botulism
	Anthrax

---

[1] For additional information on canine vaccination guidelines, see www.aahanet.org/publicdocuments/caninevaccinationguidelines.pdf
[2] Complete guidelines can be found in 2013 AAFP Feline Vaccination Advisory Panel Report. *Journal of Feline Medicine and Surgery* (2013) 15, 785–808.
[3] An adaptation of these criteria can be found at www.vetmed.ucdavis.edu

most of them include multiple pathogens (viral and bacterial). However, to date none of these have been 100% effective in preventing the disease.

As with other species, vaccine recommendations for cattle are divided into core and non-core vaccines (Table A.4). Vaccines for dairy cows and bulls are similar to those for beef cows and bulls, except that non-core vaccines also include mastitis (*Staphylococcus aureus*), hairy heel warts, and foot rot.

Because vibriosis and trichomoniasis are venereal diseases, the annual vaccine is given 30–60 days prior to breeding. Core vaccines given to calves include blackleg (7-way), IBR-BVD-PI3, and leptospirosis—all of which are administered before weaning. At 4–12 months, heifers should be vaccinated for brucellosis. Non-core vaccines for calves include BRSV, *Pasteurella*, *Histophilus somni*, pinkeye (*Moraxella bovis*), *E. coli*, anthrax, and anaplasmosis.[4]

SMALL RUMINANT (SHEEP AND GOAT) VACCINES AND VACCINATION SCHEDULES

Core vaccination recommendations for sheep and goats include one that protects against enterotoxemia, caused by *Clostridium perfringens* type C and D, and tetanus (caused by *C. tetani*). Blackleg and malignant edema vaccines, although available, are not generally recommended. Vaccination of the dam 2–4 weeks prior to parturition will provide lambs and kids with maternal immunity against the clostridial diseases (CD-T). If the dam is not vaccinated, the lambs or kids should be vaccinated at 1–3 weeks of age, with a booster 2–4 weeks later. Offspring of dams that were not vaccinated should also receive antitoxin to protect against tetanus at the time of docking, castration, and disbudding.

Non-core vaccines for sheep and goats include sore mouth (orf virus), a live vaccine that is administered to scarified skin. It is infectious to humans. The vaccine should not be used unless the herd is already infected. Other non-core vaccines include foot rot, caseous lymphadenitis, abortion agents (*Chlamydia* species, *Campylobacter fetus*), and rabies.

SWINE VACCINES AND VACCINATION SCHEDULES

Vaccines for immunization of adult swine fall into core and non-core categories, as indicated in Table A.5. All swine should receive the core vaccines because leptospirosis, parvovirus, and erysipelas are common and severe diseases against which vaccination provides protection. Vaccination of the gilt for leptospirosis before breeding can also prevent potential abortion. As with most animals, non-core vaccines are recommended depending on the prevalence of the diseases in the surrounding area.

POULTRY VACCINES AND VACCINATION SCHEDULES

There are four core vaccines that are commonly given to broiler and layer chickens (Table A.6). The use and effectiveness of the Marek's disease vaccine and the infectious bursal disease vaccine are discussed in more detail in the case studies for these two diseases (Cases 11 and 12, respectively).

---

[4] Full bovine vaccination recommendations can be found on the University of Arkansas website at uaex.edu/publications/pdf/fsa-3009.pdf

Table A.4 Core and non-core vaccines for cattle (beef cows and bulls)

Core vaccines	Non-core vaccines
Infectious bovine rhinotracheitis (Bovine herpesvirus type 2) (IBR)	Trichomoniasis
	Moraxella bovis (pinkeye)
Bovine viral diarrhea virus (BVDV)	*Clostridium* (blackleg) 7-way
Parainfluenza type 3 virus (PI3V)	Anthrax
Bovine respiratory syncytial virus (BRSV)	
Leptospirosis (5-way)	
Vibriosis	

Table A.5 Core and non-core vaccines for swine

Core vaccines (boars and gilts)	Non-core vaccines
Leptospirosis*	*Actinobacillus pleuropneumoniae*
Parvovirus	*Clostridium perfringens*
Erysipelas	Encephalomyocarditis virus
	Haemophilus parasuis
	Mycoplasma hyopneumoniae
	Porcine reproductive and respiratory syndrome (PRRS) virus
	Pseudovirus
	Rotavirus diarrhea
	Streptococcus suis
	Swine dysentery
	Swine influenza virus
	Transmissible gastroenteritis virus
	Circovirus
Core vaccines (piglets)	
Atrophic rhinitis (pre-weaning)[†]	

*Vaccination of sows and gilts before breeding is suggested to prevent abortion. Gilts receive two vaccines before breeding; sows receive one vaccine.

[†]To prevent this disease in piglets, vaccination of sows and gilts pre-farrowing is suggested in order to induce antibodies in colostrum. Also pre-farrowing vaccination against *E. coli* diarrhea is sometimes suggested to protect piglets.

Core vaccines
Marek's disease
Newcastle disease
Infectious bronchitis
Infectious bursal disease

ADVERSE RESPONSES TO VACCINATION

The majority of the commonly used core vaccines are safe for administration to the target species as indicated by the manufacturer's instructions. However, on occasion an individual may develop a hypersensitivity reaction (local and/or systemic). These reactions range from mild to severe, and can be categorized according to the Gell and Coombs classification scheme.

Local reactions are usually type I or type III. The type I reaction is mediated by IgE antibodies that are specific for one or more antigens in the vaccine. A type I reaction will occur very rapidly (within 5–30 minutes) after the vaccine is given, and can present as swelling at the injection site or localized swelling of the muzzle accompanied by pruritus. The presence of a localized type I reaction is a warning sign that subsequent vaccination could elicit a more serious systemic response.

Systemic anaphylactic reactions are the most severe type I hypersensitivity reactions. In anaphylaxis the patient will develop a shock-like condition within minutes after administration of the vaccine, due to generalized mast-cell degranulation. Clinical signs will vary depending upon the species (see Case 17, Table 17.1). Anaphylaxis is a true emergency and must be treated immediately and vigorously (with epinephrine and corticosteroids) to reverse the pathophysiology induced by mast-cell mediator liberation.

Recent research has shown that for canine and equine vaccines there is usually some contamination of the vaccine with stabilizing proteins, as well as proteins residual from cell-culture cultivation of the target virus. These substances are not injurious, but they can be antigenic (that is, some animals will respond to their presence by developing IgE antibodies). When bound to mast cells, these antibodies can elicit an anaphylactic response upon re-exposure. If several different vaccines all contain the same proteins, the response is boosted by every vaccine that contains the antigens. Fortunately, only a minority of patients develop such a response, but for those that do it can be serious or even fatal.

A type III reaction, often called the Arthus reaction, occurs at the site of injection, but unlike the type I reaction it occurs hours (typically 8–12 hours) after administration of the vaccine. The site will show an erythematous raised soft swelling, which will generally resolve within the following 2–3 days. The Arthus reaction occurs when there is pre-existing IgG antibody present in the blood and interstitial fluid. The antibody molecules bind to the antigen and create soluble complexes that stick within venule walls. At those sites they fix complement and attract neutrophils, causing local tissue damage. An Arthus reaction provides an indication that it may not have been necessary to administer the vaccine, due to the large amount of antibody already present.

The type III immune-complex reaction can also be systemic. For example, occasionally a horse that has been vaccinated with strangles vaccine will develop immune-complex disease (purpura hemorrhagica; see Case 24). In dogs, development of immune-mediated hemolytic anemia and/or thrombocytopenia is thought to be linked to vaccination, but this has not been demonstrated conclusively.

In cats, the development of injection-site sarcomas occurs in a small number of animals, but the incidence has increased since vaccination of cats for rabies and feline leukemia (both of which are killed adjuvanted vaccines) has been implemented by most feline practitioners. These tumors take years to develop, and are aggressive once discovered. Many clinics have implemented the practice of injecting both of these vaccines as low on the leg as possible (on the right for rabies; on the left for feline leukemia) so as to allow for limb amputation if necessary to save the life of the cat.

Adverse vaccine responses of all types should be reported both to the vaccine manufacturer and to the US Department of Agriculture (USDA).[5]

---

[5] Adverse reactions can be reported at
www.aphos/usda.gov/animal_health/vet_biologics/publications/adverseeventreportform.pdf

APPENDIX II
THE CLINICAL IMMUNOLOGY LABORATORY

ASSAYS TO EVALUATE INNATE IMMUNITY

The neutrophil is an important effector cell in the innate immune response. There are diseases in which these cells are either not plentiful enough or have an impaired ability to function. Recurrent bacterial infection may indicate an underlying defect in the neutrophil response. There are several points at which the neutrophil response can be affected.

PERIPHERAL BLOOD NEUTROPHIL COUNT

If the bone marrow is not producing sufficient neutrophils to enter the peripheral circulation and provide a vigorous response to infective agents, infection can go unchecked. Thus the first parameter to examine in a patient with recurrent bacterial infection is the peripheral blood neutrophil count. For example, in the gray collie syndrome (see Case 3) there is a cyclic neutropenia, which is caused by a periodic shutdown of the bone marrow that results in cycles of peripheral neutropenia. During times of neutropenia, affected dogs show an increased incidence of bacterial infection.

CHEMOTAXIS ASSAY

Neutrophils leave the blood vascular system in response to chemotactic molecules such as complement components (such as C3a). The process of neutrophil chemotaxis (movement towards the infective agent along a gradient) can be evaluated by an *in vitro* test using a Boyden chamber. In this assay the chamber has two sides separated by a barrier that is permeable to small molecules. Polymorphonuclear leukocytes (neutrophils) from the patient are placed on one side of the chamber, and a chemotactic stimulus is placed on the other side. The neutrophils will be attracted to the chemotactic gradient and will move along that gradient. The cells will pass through the porous membrane and can be detected in the lower chamber, indicating normal chemotactic ability (Figure A.1). Reduction of chemotactic ability will be manifested by random motion rather than accumulation of cells on the membrane or in the lower chamber. This translates into an ineffective response to infection in the animal. Timing of the assay is important, because eventually the gradient will dissipate.

PHAGOCYTIC INDEX

The ability of a neutrophil to engulf a pathogen and take it into a vacuole is critical for ultimate killing of the pathogen. To assess this function, neutrophils are separated from peripheral blood and are incubated with bacteria (such as *Staphylococcus* or *Streptococcus*) for 30 minutes. Cells are then harvested,

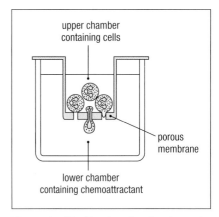

Figure A.1 The Boyden chamber is used to evaluate the ability of the neutrophil to respond to a chemotactic gradient, such as C3a, by moving towards the highest concentration of the chemotactic factor.

Figure A.2 (a) Neutrophils that have engulfed bacteria *in vitro* when incubated with the bacteria in normal serum. Opsonins from serum assist the neutrophils in binding the bacteria for engulfment. (b) Neutrophils that contain fewer bacteria. They were incubated with bacteria in a physiological salt solution with no opsonins; note that fewer bacteria have been engulfed.

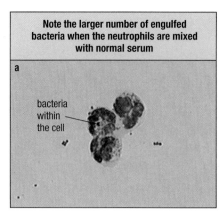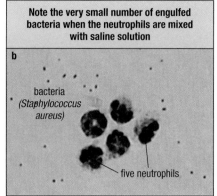

Figure A.3 Bactericidal assay in which the bacteria were added to the patient's cells and the control cells at time 0. Test time points are shown on the horizontal axis, and numbers of surviving bacteria are shown on the vertical axis. The red line shows the normal killing curve (control), and the blue line shows the lack of killing by the patient's cells.

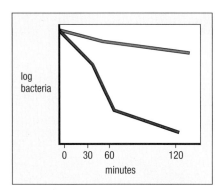

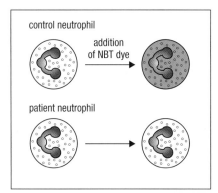

Figure A.4 The nitroblue tetrazolium (NBT) reduction test evaluates the oxidative mechanism of intracellular killing by neutrophils. Neutrophils from a normal control (normal oxidative killing) are shown in the upper part of the diagram. Neutrophils from a patient with chronic granulomatous disease are shown in the lower part. The cells of the control develop a blue-black color when the intracellular oxidative mechanisms cause reduction of the dye. The patient's cells are not able to perform oxidative killing and thus do not reduce the dye, so it remains colorless.

washed free of non-engulfed bacteria, and placed on a microscope slide, where they are stained and observed under high power. The number of bacteria engulfed is counted in 100 cells. The percentage of cells engulfed is called the phagocytic index. The ability of neutrophils to adhere to bacteria prior to engulfment is enhanced when serum containing specific antibodies to the bacteria and/or complement component C3b is present. It is therefore useful to test a patient's cells with their own serum and with serum from a normal control in order to evaluate the role of both the opsonins and the inherent phagocytic ability of the cells (Figure A.2).

INTRACELLULAR KILLING (BACTERICIDAL ASSAY)

Even when neutrophils are able to engulf bacteria, they may not be capable of killing them. To evaluate the killing function, a bactericidal test can be performed. As in the phagocytic index assay, cells are first incubated with bacteria. At intervals beginning immediately after the addition of bacteria to cells (time 0), samples are taken and treated with antibiotic to kill extracellular bacteria. Cells are then lysed and the lysate is plated onto blood agar. After incubation of the plates, colonies are counted. Typical time intervals (T) are 0, 30, 60, and 120 minutes. The results are graphed as the number of viable organisms versus the time point, and the results from the patient are compared with those from a control sample (Figure A.3). This assay only evaluates the rate of killing within the neutrophil; it does not identify the mechanism.

NITROBLUE TETRAZOLIUM REDUCTION TEST

This assay is specifically designed to evaluate the function of the myeloperoxidase system, which kills bacteria by oxidation of bacterial proteins. It is less difficult to perform than the bactericidal test, but evaluates only this particular killing mechanism. To perform the test, the neutrophils from the patient can be mixed with an inert substance such as zymosan, which is taken up in phagocytic vacuoles. The colorless nitroblue tetrazolium (NBT) dye is then added, and if the zymosan activates the killing mechanism, the dye is reduced and becomes dark blue-black in color. Cells are observed under the microscope for the presence of the dye (Figure A.4).

ASSAYS TO EVALUATE HUMORAL IMMUNITY

QUANTITATIVE IMMUNOGLOBULINS

Measurement of the amount of IgG, IgM, and IgA is a fairly simple way to determine whether the patient is making normal amounts of these

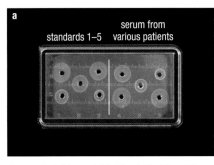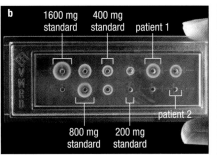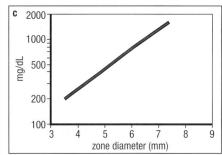

Figure A.5 (a) and (b). The left side of the plate shows the standards for a quantitative assay for canine IgA. On the right side of the plate are various patient samples. In evaluation of serum IgG levels in foals with suspected failure of passive transfer of antibody the SRID test is the gold standard with which the "quick tests" such as the SNAP® test (see ELISA) are compared. (c). An example of the type of graph that is constructed to determine the levels of immunoglobulin measured by SRID.

immunoglobulins. The assay is a single radial immunodiffusion (SRD or SRID) test in which antiserum specific for the immunoglobulin heavy chain (γ, μ, α) is incorporated into agarose in a plate. A separate test is run for each antibody class. Wells are stamped into the agarose so that samples of patient sera and standards can be placed into them. The filled wells are kept moist, and the fluid in them is allowed to diffuse into the agarose. When an area of equivalence is reached such that immune complexes of antigen (the immunoglobulin) and antibody (the antibody to the Ig heavy chain) are sufficiently large to become insoluble, a ring of precipitation is formed around the wells. The size of this ring is proportional to the amount of antibody (for example, IgG) in the well. A set of known standards are run along with the sample. By producing a graph of the ring precipitin diameter for each concentration of antibody, a standard curve is generated. The sample diameters are then compared with the standard curve, and the actual quantities of the antibody class present in the serum are determined (Figure A.5).

IMMUNOELECTROPHORESIS

This technique combines the process of electrophoretic separation of proteins with gel diffusion. It is a semi-quantitative method that is used to evaluate the presence and isotype of paraprotein in cases of multiple myeloma. It is also useful for detection of agammaglobulinemia. To perform the assay, a serum sample is placed in a well on a plate that has a thin layer of agarose gel. After electrophoresis and separation of the serum proteins, antiserum against the antibody isotypes of interest and/or whole serum is placed in troughs opposite the serum wells. Diffusion occurs, and ultimately arcs of precipitated antibody are visualized and can be stained with dye for preservation (Figure A.6).

ASSAYS TO EVALUATE CELL-MEDIATED IMMUNITY

LYMPHOCYTE STIMULATION

Evaluation of cell-mediated immunity *in vitro* can be achieved by culturing lymphocytes from a patient in 96-well tissue culture plates with mitogens. Mitogens are plant lectins that have the ability to stimulate lymphocytes to undergo blastogenesis, a process in which they incorporate nucleotides and grow larger. All T lymphocytes respond to phytohemagglutinin (PHA) and concanavalin A (Con A), and both B and T lymphocytes respond to

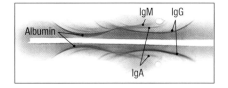

Figure A.6 The immunoelectrophoresis slide (here shown stained for preservation) shows the presence of all three major classes of antibody after incubation with antibody against whole serum in the trough. Albumin with the largest negative charge is closest to the anode, while the immunoglobulins are drawn more towards the cathode. If a serum is agammaglobulinemic, none of the antibody arcs would be present. IgE is normally present in amounts that are too small to be detected by this technique.

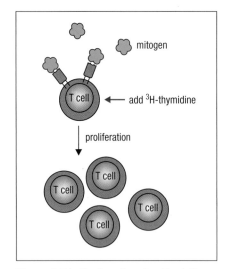

Figure A.7 In the lymphocyte stimulation test the peripheral blood lymphocytes from the patient are placed in tissue culture wells with plant mitogens. These will bind to the cells and stimulate them in a non-antigen-specific manner, which can be detected by evaluating how much tritiated thymidine (^{3}H-thymidine) is incorporated into the DNA. Visually the microscopic examination shows enlarged and dividing cells.

pokeweed mitogen. B lymphocytes also respond to bacterial lipopolysaccharide (LPS). After incubation of patient and control cells with the mitogens, they are assayed for a response. Historically this was done using uptake of a radioisotope (tritiated thymidine), which was incorporated into the DNA of the cells. The counts incorporated were measured and compared with unstimulated cells and with cells from a normal control animal (Figure A.7). More recently, a colorimetric assay has been successfully used. When stimulated cells are compared with unstimulated cells, a stimulation index is calculated. This is done for both the patient and the control animal, thus providing a means of determining whether the response of the patient is as expected.

INTERFERON γ

Another indication of a T-cell response (specifically a T-helper type 1 cell) is the production of interferon γ. This assay is performed by culturing lymphocytes from the patient with an antigen. Although many antigens can be used, the most common application of this assay is to diagnose infection with *Mycobacterium paratuberculosis* (Johne's disease) in cattle, and thus a mycobacterial antigen is used. The supernatant is harvested and the interferon γ is measured by ELISA. A kit for this assay (Bovigam®) is available. The production of interferon γ in response to the antigen is indicative of bacterial exposure or infection of the animal.

FLOW CYTOMETRY ANALYSIS OF LYMPHOCYTES

This technique can be used to determine the numbers of T cells, B cells, and T-cell subsets in populations of lymphocytes. In cases of acquired immunodeficiency, such as FIV, there will be a shift in the numbers of CD4 versus CD8 T cells. Quantitation of this ratio is useful when monitoring the progression of disease. The lymphocytes are separated from the blood using a density gradient, and are then stained with monoclonal antibody reagents that are specific for the surface marker of interest, such as CD4. A flow cytometer is then used to analyze the cell population to determine the percentage of cells staining with each antibody marker used (Figure A.8).

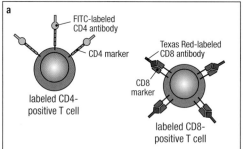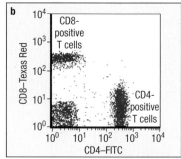

Figure A.8 (a) The labeling technique used for identification of cells bearing CD4 and/or CD8 markers. Specific antibodies are conjugated with the fluorochromes fluorescein isothiocyanate (FITC) for CD4 and Texas Red for CD8. These fluorochromes have different excitation and emission spectra, so will be able to identify the cells that they bind. The FITC produces green fluorescence, and the Texas Red produces red fluorescence. (b) A population of peripheral blood lymphocytes is treated with the labeled antibodies, and the cells staining with each are quantitated by the laser in the flow cytometer, providing percentages of cells that are either CD4 or CD8 positive. Unlabeled cells are shown in the lower left quadrant.

ASSAYS TO EVALUATE THE IMMUNE RESPONSE TO INFECTIOUS AGENTS

AGAR GEL DOUBLE IMMUNODIFFUSION (OUCHTERLONY DOUBLE DIFFUSION)

There are few diagnostic tests that use this method, but one of them has been of such great importance in the diagnosis of equine infectious anemia (EIA) that it warrants description. The Coggins test (developed by Dr. Leroy Coggins) evaluates horses for the presence of antibodies against equine infectious anemia virus. Since this virus is a lentivirus (an RNA retrovirus), once it has infected a horse it persists. A horse can be an asymptomatic carrier or can have an acute or chronic disease course. Antibodies are made against the virus but are unable to clear it. To perform the Coggins test, agar is poured into a small dish and allowed to harden. Wells are then stamped into the agar in the shape of a hexagon, with one central well. EIA viral antigen is placed in the center well, and sera from test horses are placed in the surrounding wells. Sera from positive control horses alternate with sera from patients. After the antigen and serum have diffused into the agar and formed immune complexes, the results are read by observing whether a line of precipitation is apparent between serum and antigen wells and, of most importance, whether or not the line shows an identity reaction with the positive control serum on either side (Figure A.9). The agar gel double immunodiffusion test is not nearly as sensitive as primary binding tests such as ELISA, but it works extremely well for the diagnosis of EIA in horses, for which it is considered to be the "gold standard."

AGGLUTINATION

When the antigen of interest is particulate (for example, red blood cells or bacteria), one way to test for antibodies against it is to look for clumping or agglutination in the presence of serum. To determine an antibody titer to a bacterial pathogen, such as *Salmonella*, the serum is diluted (usually two-fold) in test tubes, and an equal amount of an inactivated preparation of the bacterium is added to each tube. After incubation the tubes are evaluated for agglutination (spreading of the cells over the bottom of the tube) versus simple settling of the bacteria by gravity (Figure A.10). The amount of agglutination is graded on a scale of 1+ to 4+, and the last tube with 2+ agglutination is usually regarded as the endpoint. Thus the titer of the serum is the inverse of the dilution in that tube.

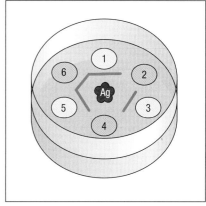

Figure A.9 The Coggins test with three positive control sera interspersed between test sera from three horses. Viral antigen (Ag) is in the center well. Horse number 6 is positive because it is showing a line of identity between the positive controls on either side. The other two horses (numbers 2 and 4) are negative for antibodies to the virus and therefore negative for EIA.

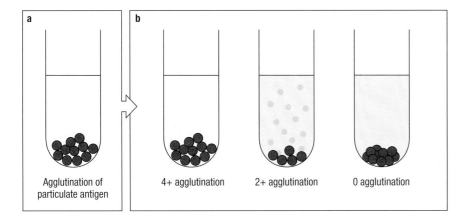

Figure A.10 (a) Agglutination reaction on the bottom of the tube. (b) Comparison of 4+ agglutination (in which all of the bacteria are removed from suspension), 2+ agglutination (in which 50% of the bacteria remain in suspension), and 0 agglutination (a negative test in which the bacteria settle by gravity, and some bacteria remain in suspension due to the lack of antibody to agglutinate them).

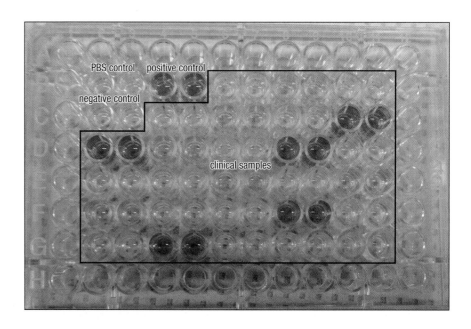

Figure A.11 A 96-well plate used for quantitative ELISA. When used to detect antigen, specific antibody is coated onto the wells. When used to detect antibody, the antigen of interest is coated onto the wells. In the photo, horseradish peroxidase-conjugated antibody is used to show a positive well (dark color). PBS is a buffered saline substituted for sample to control for any nonspecific binding of reagents. Samples are always run in duplicate or triplicate, and the mean value is used to calculate the result. The optical density of the wells can be read on an ELISA reader.

THE COMPLEMENT FIXATION TEST

This type of test has historically been very important in checking for infection caused by a variety of pathogens, viruses, protozoa, and bacteria. It is a difficult test to run and requires many controls. For this reason it has been largely replaced by other assay types, such as ELISA. The principle of the complement fixation test is competition between one system (serum being tested for antibodies against an antigen) and an indicator system composed of sheep erythrocytes and anti-sheep erythrocytes. Enough complement is added to react with only one system. If there are no antibodies against the antigen in the first system, the complement lyses the sheep erythrocytes in the indicator system. When performed with serum dilutions in test tubes, this test can be used to establish an antibody titer.

ENZYME-LINKED IMMUNOSORBENT ASSAY (ELISA)

This assay is a very sensitive and versatile platform for determination of either antibody titers or the presence of antigen. ELISA can be quantitative, or can simply give a positive or negative result. The test is based on the binding of an antibody that has been conjugated with an enzyme to antigen. After washing steps to remove unbound antibody, a substrate is added; if the enzyme is attached it will act on the substrate and cause a color change. A solid substrate is used, but this can range from wells of a 96-well microtiter plate to small plastic devices. If an ELISA test is to be quantitative, it is generally performed in a microtiter plate and the optical density of wells is measured as shown in Figure A.11.

The SNAP® Foal IgG Test is semi-quantitative, as it places the sample in a range of IgG concentrations. Figure A.12 shows a previous version of this test called the CITE® test. This type of test is semi-quantitative and provides enough information to enable the veterinarian to determine whether sufficient blood levels of IgG have been established in the foal.

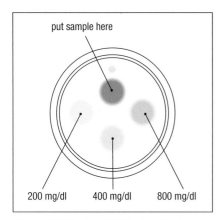

Figure A.12 In the CITE test, patient serum is compared with dots that have various amounts of equine IgG affixed to the membrane. Here the sample indicates that the foal has over 800 mg/dL and therefore does not have failure of passive transfer.

When antigen is to be detected, as in the test for feline leukemia virus (FeLV), there is antibody against the virus on the solid substrate. If antibody is to be detected, as in the test for feline immunodeficiency virus (FIV), antigen is present on the solid substrate. Positive and negative controls are always included. SNAP® tests for FeLV and FIV are for antigen and antibody, respectively, and are not quantitative, but this does not present a problem, as the tests are used to determine whether or not the cat is currently infected with the virus. There are many different commercially available kits for determining whether a cat is FeLV positive or negative. Many of them use a solid format that is easy to implement in an office setting.

IMMUNOFLUORESCENCE

Immunofluorescence can be used in a variety of ways. The technique utilizes the fact that fluorochromes can be excited by distinct light waves and emit at different frequencies. When fluorochromes are conjugated to antibodies they can be used to detect antigen *in situ* or on cell surfaces. For determination of antibody titers to some viruses or other intracellular pathogens, serum is incubated with permeabilized cells on slides, and antibodies in serum are allowed to bind. The slides are then washed and a secondary antibody (conjugated to a fluorochrome such as fluorescein) is added. After incubation and washing, the slide is examined under a microscope equipped with a mercury vapor lamp, special filters, and dichroic mirrors to detect the fluorescence. Cells with intracellular virus that has bound antibody from the test serum will show intracellular fluorescence. Using serum dilutions, this technique can be used to determine titers (indirect ELISA). The direct fluorescent antibody (FA) technique can be used to detect antigen from pathogens within tissue sections (for example, distemper virus in a conjunctival smear from a dog, or equine herpesvirus in equine lung cells).

IMMUNOHISTOCHEMISTRY

This technique can be used to detect antigen from pathogens within tissue sections. In principle it is quite similar to ELISA, and uses enzyme conjugated to antibody to detect antigen. However, it can detect antigen *in situ*, in tissue sections, and is thus an excellent way for the pathologist to confirm that a particular virus is present in an organ.

HEMAGGLUTINATION INHIBITION (HI) TEST

Some viruses are able to agglutinate erythrocytes of some species using a hemagglutinin molecule found on the surface of the virion. When antibodies are produced that react with these molecules, they bind to the hemagglutinins of the virions and prevent them from binding to the erythrocytes, thus preventing the agglutination of the erythrocytes. This attribute of these viruses has been used to develop an assay to measure the presence and amount of antibody specific for the virus. To perform this assay, a serum sample is serially diluted, and a standard aliquot of erythrocytes and virus is then added. After incubation, the test is read for agglutination. The negative control, which lacks anti-hemagglutinin antibodies, will show agglutination of the erythrocytes by the virus, whereas the positive serum will show no agglutination of the erythrocytes.

ASSAYS TO EVALUATE AUTOIMMUNE AND HYPERSENSITIVITY RESPONSES

ANTINUCLEAR ANTIBODY (ANA) TEST

In the diagnosis of systemic lupus erythematosus, and certain other autoimmune diseases in humans, the presence of a positive ANA test is an important diagnostic criterion. The test detects the presence of antibodies that are able to react with various nuclear components (for example, single-stranded DNA, double-stranded DNA, and histones). The assay is performed as an indirect immunofluorescence test. A cell line called HEp-2 is grown on slides with wells and made permeable by acetone treatment. The patient's serum is

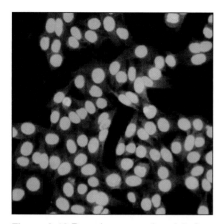

Figure A.13 Round green oval nuclei of the HEp-2 cells are fluorescent because the patient's serum contains antibodies against one or more of the internal components of the nucleus.

diluted from 1:20, and dilutions are added to the cells on the slide. After incubation and washing, a fluorescein-conjugated antibody against the patient's antibody (species specific) is added and incubated. After a final wash the slide is examined for the presence of nuclear fluorescence. Positive and negative controls on each slide are used for comparison. The titer is the last well that shows the presence of nuclear fluorescence (Figure A.13). Various different patterns of nuclear fluorescence are recognized; most common in the dog are the speckled and homogeneous patterns, but on occasion a peripheral ring or a nucleolar pattern is seen.

ANTI-MEGAKARYOCYTE ANTIBODY TEST

This test is used to determine whether there is an autoimmune response against cells in the bone marrow that make platelets (thrombocytes). Low levels of platelets in the peripheral blood are sometimes the result of autoimmune thrombocytopenia. A bone-marrow aspirate is obtained and stained with fluorescein-conjugated antibody against the patient's IgG. In this immunofluorescence assay, megakaryocytes are examined for the presence of fluorescence, indicating that the patient's antibodies have bound.

MAJOR AND MINOR BLOOD CROSSMATCHING

When a blood transfusion is required, donor blood and recipient blood are tested using crossmatching, which is a type of agglutination test that is performed to evaluate the compatibility of blood for a potential transfusion. In the major crossmatch the erythrocytes from the potential blood donor are mixed with the plasma of the recipient. Agglutination indicates that the patient has antibodies that will react with the cells of the donor. In the minor cross match, the erythrocytes of the patient to be transfused are mixed with the plasma of the potential donor to evaluate the presence of agglutinating antibodies.

COOMBS TEST

Also called an anti-globulin test, the Coombs test is performed to determine whether a patient has autoantibodies against erythrocytes. In the direct Coombs test the patient's red blood cells are mixed with antiserum against IgG (species specific), and the reaction is read for agglutination. If there is antibody bound to the red blood cells, the anti-IgG will cause the cells to agglutinate. If there is no bound antibody, the cells will fail to agglutinate. In the indirect Coombs test the serum from the patient is mixed with normal red blood cells from the same species. If there is circulating antibody present it will bind to the red blood cells, and then when the anti-IgG reagent is added agglutination will occur; otherwise there will be no agglutination (Figure A.14).

DIRECT IMMUNOFLUORESCENCE ON TISSUE BIOPSY

This is a versatile technique that can be used to demonstrate the presence of immune-complex deposition in tissues, such as kidney glomeruli (Figure A.15), or in skin sections from patients with autoimmune disease, such as pemphigus. An antibody reagent conjugated with fluorescein and specific for either antibody (IgG) or C3 (complement component deposited with immune complex) from the species that is being evaluated is added to the tissue section. After incubation and washing steps, the tissue is observed under a microscope equipped with epi-illumination for fluorescence, and immune complexes are visualized if present.

POLYMERASE CHAIN REACTION (PCR)

The PCR is not an immunological test—it is a technique that can be used to amplify a particular gene in a DNA sample. PCR utilizes primers which are made up of nucleotide sequences that frame the area of interest (3′ and 5′

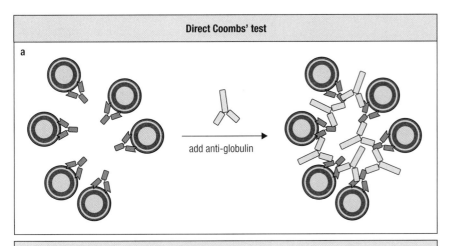

Direct Coombs' test

a

add anti-globulin

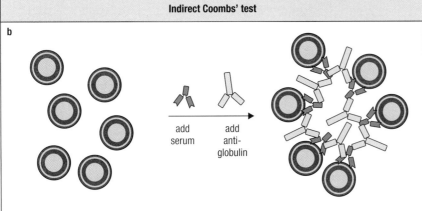

Indirect Coombs' test

b

add serum   add anti-globulin

Figure A.14 (a) In the direct Coombs test, antibody (shown in blue) that is already bound to the patient's erythrocytes is detected using an anti-globulin (Coombs reagent). If there is antibody on the cells, the anti-globulin (shown in green) will bind and cause the erythrocytes to agglutinate. If there is no antibody on the cells, they will not be agglutinated. (b) In the indirect Coombs' test, the presence of erythrocyte-binding antibodies in the serum of the patient is detected by using normal erythrocytes of the same species and incubating them with the patient's serum. If anti-erythrocyte antibodies (shown in blue) are present they will bind the erythrocytes, and when the anti-globulin reagent (shown in green) is added agglutination will occur. A control consisting of the erythrocytes and anti-globulin serum in the absence of the patient's serum will fail to agglutinate.

ends). In the context of the cases presented in this book, PCR is used to identify a mutation that prevents formation of a biologically active CD18 molecule, causing the BLAD syndrome in cattle. It is also used to identify Arabian horses that possess the mutation for the defective p350 that makes the DNA-dependent protein kinase ineffective, thus preventing DNA repair in the SCID foal. PCR can also be used in conjunction with a reverse transcriptase step to convert RNA to complementary DNA prior to performing PCR. For diagnosis of some viral diseases, such as BVDV infection, an ear notch is used for detection of the viral genome in the tissue of the bovine patient.

VIRUS NEUTRALIZATION TEST

Although there are many assays for measuring antibodies against viruses, most of them do not distinguish between antibodies that simply bind to the virus and antibodies that actually impair the ability of the virus to infect cells. The virus neutralization test (also called serum virus neutralization test) is an assay designed to measure the titer of virus-specific antibodies that are able to bind to surface molecules on the virion and prevent it from infecting cells. This type of assay is an excellent correlate of protection. To perform the assay, the permissive cells that support viral replication are grown in tissue culture wells. Triplicate wells are then infected with the virus (positive infection control) or treated with media only (negative infection control). Other well triplicates are infected with aliquots of the same virus that has been mixed with a dilution of the patient's serum. After several days of growth, the cells are observed for cytopathic effect (CPE)—that is, for evidence of sick and dying cells. A titer is determined by looking for the last dilution of serum that protects the cells from CPE.

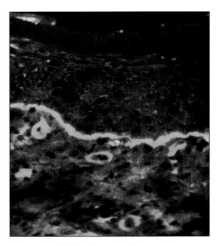

Figure A.15 This is a skin biopsy from a dog with the autoimmune disease systemic lupus erythematosus (SLE.) It has been stained with anti-IgG-FITC conjugate. Observation under the epi-illuminated microscope shows the presence of a bright band at the dermal-epidermal junction where immune complex deposition has occurred. In addition, several mononuclear cells in the dermis show cytoplasmic fluorescence. This same technique is often used to delineate the deposition of immune complexes in kidney glomeruli.

ANSWERS

CASE 1

1. Other innate immune mechanisms that protect the lungs include the turbinate bones in the nose, which entrap larger particles before they can reach the trachea, and the alveolar macrophages, which pick up particles and bacteria within the lung. The ability of these cells to move material out to the nasopharynx is reduced in cases of primary ciliary dyskinesia.

2. Other ciliated structures, such as the inner ear and the reproductive tract, often show abnormal function in patients with ciliary dyskinesia. These defects have the potential to affect both hearing and fertility.

3. If Angel had been bred to a dog that was a carrier of the gene, in a litter of four puppies one would expect two puppies to be affected (double recessive) and two puppies to be carriers (heterozygous for the defect).

CASE 2

1. The absence or decrease of a functional CD18 due to the defective β2 integrin (CD11/CD18) impaired the ability of neutrophils to bind to vascular endothelial cells. Therefore neutrophils were unable to exit the vessels and accumulate at areas of infection. The bone marrow responded to the infection by increasing production of neutrophils, accounting for the increase in numbers of circulating neutrophils.

2. Antibodies have an important role as opsonins. They bind the bacteria with specific Fab, and also bind to the neutrophil phagocyte through binding of the antibody Fc portion to Fc receptors on the neutrophil membrane. These antibodies thereby enhance phagocytosis. Although antibodies can agglutinate bacteria and fix complement, without functional neutrophils at the site of bacterial infection the role of antibodies as opsonins is limited.

3. Expression of the BLAD trait requires the presence of two mutant recessive genes. The genetic combination that results in a homozygous recessive calf occurs when the bull and cow are both carriers of the recessive mutant gene (heterozygous). Neither carrier is clinically affected. In such a mating there is a 25% chance of the calf inheriting both recessive mutant genes and thus expressing the BLAD trait. A very popular Holstein bull (Ivanhoe) was a heterozygous carrier. Widespread utilization of his semen for artificial insemination created numerous carriers of the gene, and thus increased the likelihood of breeding two heterozygous cattle.

CASE 3

1. Neutrophils have a very short life in the periphery, so any decrease in influx of new neutrophils from the bone marrow is recognized rapidly. The neutrophil is also the first line of defense against many pathogens, particularly bacteria. Without the innate recognition and killing of bacteria by polymorphonuclear neutrophils, bacteria can replicate unchecked. Once antibodies are produced they can control most bacteria, but the results of initiation of the humeral response take days, not hours, to become effective.

2. Although the cycle timing is usually different from that of the neutrophil, production of other bone-marrow cells also fluctuates in this disease. One of these cells, the platelet (thrombocyte), is responsible for blood clotting. Periods when there are low numbers of platelets can result in excessive bleeding if a wound is incurred.

3. A bacterial infection such as pneumonia normally stimulates the production of neutrophils and the liberation of many more of them than usual into the blood, causing a peripheral neutrophilia, which is observed on a complete blood count (CBC). In a patient with cyclic neutropenia the acute bacterial infection occurs because there are insufficient numbers of neutrophils, and their absence is apparent on the CBC.

CASE 4

1. All three cases involve a deficiency in the ability of neutrophils to protect the host, but each one involves a different aspect of neutrophil biology and thus a different type of neutrophil impairment. The calf with BLAD lacks a functional integrin on the neutrophil surface, and the neutrophil is consequently unable to marginate and move out of the blood vascular system. The gray collie with cyclic neutropenia has normal neutrophils that are capable of chemotaxis and killing of bacteria, but lacks sufficient neutrophils during cyclic periods of bone-marrow suppression. In this case (a cat with Chediak–Higashi syndrome), because the neutrophils have abnormal lysosomes they are physically unable to kill bacteria.

2. The mutation that causes the syndrome affects both melanin pigment in hair shafts and granules in the neutrophil. The beige mouse has a mutation in the *LYST* gene (lysosomal trafficking regulator). The LYST mutation causes hypopigmentation due to inappropriate fusion of premelanocytes with lysosomes. The result of this mutation is the formation of huge granules in melanocytes, granulocytes, mast cells, and retinal pigment cells. These cells, which include natural killer (NK) cells, are unable to carry out their normal function because of these very large lysosomal granules. The beige mouse version is almost identical to the human version of Chediak–Higashi syndrome, as well as that of the cat and Brangus cattle.

3. Bone-marrow transplantation would bring in fresh myeloid progenitor cells without the CHS1/LYST mutation. Cells that developed from these progenitors would be functional.

CASE 5

1. IgM is made by the foal and is not obtained through the colostrum. By the age of 6 weeks there should be a significant amount of detectable IgM in the circulation. IgG is obtained through the colostrum. A young foal can have both IgG that it has made and IgG that is still present as a result of passive absorption from the colostrum. The colostral antibody protects the foal

during the early weeks of life, but it decays over time, and by 8 weeks of age its levels are generally very low. Thus in the foal with SCID the IgG present is of maternal origin.

2. The DNA-dependent protein kinase enzyme is responsible for repairing the break that occurs in the DNA after the unwanted *V*, *J*, and *D* genes have been "looped out" of the DNA. To create functional T- and B-cell receptors, the ends of the DNA must be repaired by this enzyme. If the enzyme is not functional, the repair cannot occur and there will be no receptors. Without the receptors, lymphocytes will undergo apoptosis and thus will not survive to populate the lymphoid organs.

3. Mendelian genetics dictate that when a trait is autosomal recessive the possible combinations of alleles from two heterozygotes provide the following ratio: one homozygote for the unaffected gene, two heterozygotes for the carrier status, and one homozygote for the defective gene. Thus the probability of obtaining a SCID foal as a result of breeding two carriers is 25%. Since the trait is autosomal recessive there is no sex link, and the defect is expressed equally in male and female offspring.

4. Yes. As horses can have their DNA tested for the mutation, it is possible to prevent affected offspring by not breeding two carrier animals together. The test that is used to determine whether the mutation is present is polymerase chain reaction (PCR).

CASE 6

1. The mechanism that causes selective IgA deficiency is not known. However, there are several points during the immune response at which an error might result in IgA deficiency. During B-cell development, the change from production of IgM or IgG to production of IgA requires that there is an intact alpha DNA exon, and that the switch region immediately preceding it is functional. Mutations in either of these genes could prevent differentiation of B cells into plasma cells that produce IgA. The T-helper type 2 lymphocyte that produces IL-10 and IL-5 has been shown to facilitate the switch to IgA. Thus a deficiency in the production of these cytokines may be a causative factor. Since both serum and secretory IgA were deficient in this dog, it is unlikely that the defect was related to the secretory mechanism. If serum IgA was normal and secretory IgA was low or absent, structural or functional abnormalities in the polymeric immunoglobulin receptor would be a potential site for a defect.

2. Both human and canine IgA-deficient patients may develop infections in the ear canal due to overgrowth of bacteria and/or yeasts. The lack of IgA in the secretions within the sebum of the ear allows bacteria to grow and to adhere to aural cells, which would normally be afforded some protection by the presence of these antibodies.

3. An increased incidence of respiratory allergy has been reported in both human and canine patients with selective IgA deficiency. IgA in airway secretions can bind to and remove inhaled pollens and other allergens before they are able to gain access to antigen-presenting cells and mast cells in the sub-epithelial lamina propria, thereby preventing sensitization and/or elicitation of an IgE-mediated response. If IgA is absent, the allergens have greater access to immunologically active cells, and stimulation of the IgE response is enhanced, particularly in those individuals that are genetically high IgE responders.

4. The intranasal vaccine for kennel cough is a modified live/attenuated vaccine that is administered onto the respiratory mucosa to stimulate a secretory IgA response. Therefore a dog with selective IgA deficiency should not

be vaccinated by this route. The parenteral vaccine elicits an IgG response. This IgG will be present in blood and will find its way into secretions, albeit at lower concentrations and less consistently than IgA. The IgG response will therefore be protective in the IgA-deficient patient.

CASE 7

1. The 2-day-old foal has already undergone intestinal "closure" (by 24 hours of age; see Figure 7.1), so any IgG consumed at that time will not be taken up into the circulation, but will remain in the intestine and be digested as a protein source.

2. The human infant (or primate) obtains all of the IgG required for passive protection via placental transfer from the maternal circulation during gestation. Thus consumption of colostrum by the neonate may add supplemental sources of IgA that are active in the intestine, but the neonate will be protected without colostrum by the prenatally transferred IgG.

3. Sometimes a mare has poor-quality colostrum due to poor condition and/or nutrition, and despite consuming what should be a sufficient amount of colostrum her foal may consequently have a lower than adequate serum level of IgG.

CASE 8

1. The latency of the virus allows the infected cells to persist undetected because they do not express viral RNA or protein. To clear the retroviral infection, FeLV-infected cats would need to eliminate all of the infected cells; this scenario is unlikely.

2. FeLV can induce cellular transformation by (a) integration adjacent to a cellular proto-oncogene, (b) integration-induced disruption of a cellular tumor suppressor gene, or (c) formation of a chimeric, replication-defective feline sarcoma virus that carries viral oncogenes.

3. The FeLV ELISA is an assay for viral antigen, whereas the FIV ELISA is an assay for viral antibody. As a result, cats vaccinated for FeLV (with detectable antibody but no viral antigen) will not be mistaken for infected animals. Cats vaccinated for FIV will also have detectable viral antibody. The standard FIV ELISA Snap test is not able to distinguish between virus-induced antibody and this vaccine-induced antibody. However, several newer tests that utilize the P40 viral protein as antigen (Witness and Anigen) are better able to distinguish Vaccine from Infection antibodies.

CASE 9

1. The nature of the lentivirus is to be incorporated into the genetic material of the cat. Antibodies are produced, but they do not eliminate the virus. Thus the presence of antibodies is indicative of infection.

2. In addition to CD4+ T lymphocytes, FIV is able to infect dendritic cells, which are critically important in innate immune responses as they act as sentinel cells for the body. For example, antigen processing and presentation are functions performed by dendritic cells that activate both innate and acquired immune responses. The disruption of dendritic cell function by FIV therefore leads to suppression of innate immune responses.

3. Whenever there are simultaneously high levels of circulating antibody and circulating antigen, immune complexes form. In the case of the FIV-infected cat, there is chronic stimulation of the immune system causing production of antibodies to FIV. These antibodies can precipitate glomerulonephritis via immune complexes forming and lodging in kidney glomeruli, stimulating a type III hypersensitivity response. Subsequent neutrophil chemotaxis and degranulation causes tissue destruction and compromises glomerular function.

4. Like FIV, FeLV is classified as a retrovirus. However, unlike FIV-infected cats, cats that test positive for FeLV may recover and stay healthy, or can become persistently infected with FeLV and die from FeLV-related disease. Cats that test positive and then clear the infection will test negative within a few months. This is an important distinction between the two viral infections: after the primary FeLV viremia, many cats eliminate the virus, whereas this does not happen with FIV. Those FeLV-infected cats that fail to eliminate the virus undergo a secondary viremia and generally remain infected. Lymphoma and anemia are common in FeLV-infected cats. As with FIV infection, FeLV weakens the immune system and allows secondary infections to proliferate. However, FeLV does not target the CD4⁺ T lymphocytes, and thus has a different mechanism of immune suppression. The test for FeLV is for antigen in the blood or saliva. FeLV can cause some of the same clinical signs as FIV infection, such as anorexia, stomatitis, diarrhea, persistent fever, and neurological disorders. FeLV is transmitted in saliva, nasal secretions, and from the milk of an infected mother; infection of kittens is common. Like FIV, FeLV can also be transmitted via bite wounds.

CASE 10

1. Lymphoid tissues are targeted by BVDV. At necropsy, depletion of lymphoid follicles in the gut and associated lymph nodes is observed. BVDV can enter lymphocytes and macrophages, where it can induce apoptosis of lymphocytes and compromise the phagocytic functions of macrophages.

2. Infection with BVDV prior to completion of positive and negative thymocyte selection causes BVDV antigens to be recognized as self. This allows the virus to persistently infect the calf without antibody or cytotoxic T-cell recognition. Virus can be shed throughout the life of the persistently infected animal.

3. The ear notch is taken from the calf and tested for BVDV antigen by antigen-capture ELISA. It is also possible to test pooled ear notch samples from a herd by PCR. The presence of the viral antigen is indicative of infection. BVDV antigen persists in the skin.

CASE 11

1. At this point there is very little that the owner can do. If the birds were vaccinated *in ovo* or on day 1 with one of the cell-associated vaccines, they are likely to have some protection. The chickens are latently infected by virus present in the environment. The incidence of disease decreases dramatically after approximately 20 weeks of age.

2. They should contact the hatchery or feed store and make sure that the hatchery vaccinates against Marek's disease *in ovo* or on day 1 with one of the cell-associated vaccines. The chicken house should also be thoroughly cleaned and disinfected before new birds are brought in.

3. Marek's disease virus (MDV) preferentially infects cells of the feather follicle, and can remain viable in feather dander for several months. The viable virus can be inhaled by susceptible chickens from desquamated epithelium in poultry-house dust shed from feather follicle epithelium, and can cause infection of B cells and ultimately transformation of T cells and lymphoproliferation.

4. If the hens were vaccinated, neutralization of maternal antibodies in the chicks may occur and could result in an ineffective vaccination. Maternal antibodies persist for up to 3 weeks in chicks. Vaccination after maternal antibody has waned would increase the probability of achieving effective immunization.

CASE 12

1. In avian species the bursa of Fabricius is a primary lymphoid organ; it is the site of maturation of B lymphocytes. The precursors come from the bone marrow, and in the bursa they develop the appropriate cell markers and receptors that will allow the B lymphocytes to respond to antigens and to leave the bursa to seed the blood and secondary lymphoid organs with B cells. Ultimately these B cells will be the source of antibody production for the bird. Neonatal bursectomy will result in a bird with few or no B cells (depending on the timing of bursectomy).

2. Infectious bursal disease virus (IBDV) causes an inability or decreased ability to produce antibodies against pathogens, making the bird very susceptible to additional bacterial or viral infections.

3. A bird with immunosuppression due to IBDV infection would not respond well to antigenic challenges such as vaccination because the decreased B-cell function would limit antibody production in response to the vaccine antigen.

CASE 13

1. Respiratory viruses such as bovine respiratory syncytial virus (BRSV), infectious bovine rhinotracheitis (IBR), and parainfluenza type 3 virus can increase the likelihood of bacterial infection of the bovine lung in several ways. BRSV can synergize with *Histophilus somni* to increase pro-inflammatory cytokines, prostaglandin E2, and matrix metalloprotease production by respiratory epithelial cells, thereby enhancing the environment for bacterial proliferation. In addition, by destruction of ciliated epithelial cells, the normal mechanisms that prevent bacteria from traveling from the upper respiratory tract into the lung are compromised. In the case of BVDV the virus is immunosuppressive, causing depletion of lymphoid cells in follicles in the gut and in lymph nodes.

2. The stress associated with weaning, castration, and shipping causes the release of cortisol from the adrenal gland. Cortisol has anti-inflammatory effects that include depression of neutrophil chemotaxis and killing. Among other negative effects on the immune system, cortisol can decrease IgA secretion in the lung and prevent T cells from responding to interleukin 2, thus limiting T-cell clonal expansion.

3. Serological tests can be performed to determine the viral etiology. The most likely viruses would be BRSV, IBR, parainfluenza type 3 virus, and possibly bovine viral diarrhea virus (BVDV). Titers can be obtained by indirect immunofluorescence, ELISA, or virus neutralization. If animals have been previously vaccinated, determination of the causal virus would require both

acute and convalescent titers and/or virus isolation, or reverse-transcription polymerase chain reaction (RT-PCR) during the acute phase.

CASE 14

1. It is most likely that Smokey developed difficulty breathing and a cough while living inside because her immune system responded to house-dust mites and/or pollens brought into the home through open windows or on the clothing of her owners. The IgE sensitized her mast cells by binding to Fc-epsilon receptors. When she was re-exposed to the allergen(s), the IgE bound them and triggered the release of mediators from the mast cell. Histamine causes smooth muscle contraction and increased secretions; serotonin is also an important mediator in the cat. In addition, the synthesis of leukotrienes is triggered, which would further exacerbate the airway constriction and would trigger the migration of inflammatory leukocytes (including eosinophils) to the area. The allergens that are most likely to have induced asthma in this case include house-dust mites and aeroallergens (pollens brought inside on clothing or through open windows).

2. The eosinophilia was stimulated by several mechanisms. Release of IL-5 from T-helper type 2 lymphocytes causes the bone marrow to increase the production and release of eosinophils into the periphery. Once the mast cell has been activated, additional chemokines, such as CCL-11 (eotaxin), can attract and activate eosinophils. The eosinophils follow the chemotactic gradient to sites of immune reactivity in the connective tissues in and around the airway basement membranes.

3. If Smokey's owner had opted for skin testing or serum IgE testing, the allergens could have been identified. Once they had been identified, the veterinarian and Smokey's owner could have decided on a therapy, which might include avoidance of the allergens and/or specific immunotherapy.

4. The inhaled corticosteroid decreased inflammation in the lung and, taken on a regular basis, prevented the accumulation of inflammatory cells. Albuterol is a beta-2 adrenergic agonist. It stimulates beta-adrenergic receptors. Binding of albuterol to beta-2 receptors in the lungs causes the bronchial smooth muscle to relax, thus making it easier for the cat to breathe.

CASE 15

1. The mast cell is an important effector cell. It binds IgE with specificity for the allergen, and upon binding of that IgE with allergen it is triggered to undergo granule exocytosis. The mast-cell mediators have an important role in causing the physiological effects that are manifested in the skin. The mast cell also releases cytokines, such as IL-4, which will further enhance development of the allergic phenotype.

2. A test that is positive for serum allergen-specific IgE means that there is circulating IgE in the blood at the time of sampling. A positive intradermal skin test for IgE means that there is allergen-specific IgE present on the mast cells. These results are generally in agreement when the allergen is currently in the environment to further stimulate the IgE response. Since the half-life of IgE is about 2 days, the IgE levels will fall quickly once the stimulus has been removed. The IgE on the mast cell remains attached to the FcεR1 for many months (very-high-affinity binding), and thus the result of an intradermal test can remain positive even in the absence of continual stimulation from allergen in the environment. Therefore for house-dust-mite allergen the test results will probably be in agreement, but for a seasonal

allergen, such as a grass pollen, the serum test result may be much lower than the skin test result.

3. Prednisone is a synthetic glucocorticoid receptor agonist that binds to glucocorticoid receptors and suppresses inflammation in several ways. After being metabolized to prednisolone in the liver, it causes inhibition of leukocyte infiltration at the site of inflammation, depresses the function of inflammatory mediators, and at high doses can suppress antibody production. Prednisone also inhibits *COX-2* gene transcription, which can result in depressed levels of prostaglandins. Atopica® (cyclosporin capsule) is an immunosuppressive drug that inhibits T-cell function by depressing interleukin 2 production. Apoquel® (oclacitinib tablet) targets pruritus and inflammation by interfering with the Janus kinase pathway (JAK1); this pathway is responsible for the itching associated with atopic dermatitis.

CASE 16

1. The hypersensitivity reaction is localized to the sites where salivary antigen from the *Culicoides* gnat enters the horse's skin to sensitize and later to elicit mast-cell degranulation, causing histamine release with resultant pruritus. Lesions therefore occur at the site of *Culicoides* feeding. The most common feeding sites are the inner part of the pinnae, the dorsal neck/base of mane hair, the tail dock, and sometimes the ventral midline.

2. The mechanism of IgE production is the same whether the allergen is injected (as in an insect bite) or inhaled (as in asthma). The response starts with antigen binding to antigen-presenting cells (APC), Langerhans cells in the skin, and dendritic cells in the lung. The cytokine environment stimulates T cells to become T-helper type 2 cells. The binding by T-cell receptors to antigenic peptides presented on MHC class II molecules (as well as co-receptor molecule binding by CD4+ T cells) stimulates B-cell proliferation and differentiation into plasma cells. These plasma cells produce IgE specific to the antigens/allergens that initiated the response. Once it has been produced, the IgE binds tightly to mast-cell Fc-epsilon receptors (type 1), and the horse is ready for elicitation of the allergic response the next time the allergen is introduced. In horses with *Culicoides* hypersensitivity the response will occur within the skin. In contrast, in the asthmatic cat the mast cells sensitized with IgE are in the respiratory tract. When allergen is inhaled the allergen cross-links the IgE (as it does in the horse skin) and mediators are liberated. In the case of the asthmatic cat the histamine stimulates contraction of bronchial smooth muscle, and dyspnea occurs as a result. In horse skin these same mediators cause increased capillary permeability and chemotaxis of leukocytes to cause inflammation.

3. Researchers are not sure why there is not always a clear correlation between skin test positivity and induction of the allergic clinical signs in the horse. One possible explanation may be the presence of high levels of IgG in interstitial tissue in the skin, which act to "mop up" introduced allergen before it can bind to IgE on mast cells (thereby preventing clinical signs in an IgE-positive horse). Another possibility is that some of the more crude preparations of allergens can elicit mast-cell degranulation without the presence of IgE on mast cells (perhaps by eliciting complement factors C3a and C5a, which can degranulate mast cells). The allergen-specific IgE ELISA detects circulating IgE, and its use in conjunction with skin testing may be a superior way to differentiate real from false-positive skin tests.

CASE 17

1. In Jet's case, her close exposure to chickens and their excrement for many years probably sensitized her through the respiratory tract. Dendritic cells in the lung processed these avian antigens and presented them to T-helper cells. The cytokines interleukin 4 and interleukin 13 were secreted to stimulate B lymphocytes to differentiate into plasma cells that produced IgE specific for the avian proteins. Mast cells throughout the body and especially in the lung bound the IgE, and the horse was effectively sensitized. When the vaccine was injected, the avian proteins from the egg embryo preparation were rapidly taken up into the blood and distributed throughout Jet's body. Mast cells in the lung and gut degranulated and released the mediators (histamine and serotonin) that elicited the clinical signs.

2. The wheal that occurred the previous year was a local anaphylactic reaction. In the area of vaccine deposition, mast cells with IgE on their surface degranulated. Histamine produced increased vascular permeability, causing fluid to leak from the blood vessels and a wheal to develop. The wheal development was a warning sign that Jet might show a more severe response in the future.

3. Epinephrine acts on $\alpha1$ and $\beta2$ adrenergic receptors to alter the physiological effects of the mast-cell mediators in anaphylaxis. Acting as a beta agonist, it dilates smooth muscle, helping to open airways that constrict in response to histamine and other mediators. Acting on alpha receptors, epinephrine has a vasoconstrictive effect and counteracts the vasodilation produced by histamine. These effects of epinephrine enable the patient to breathe and re-establish the homeostasis that is required to maintain blood pressure.

CASE 18

1. In equines, antibodies in colostrum are absorbed intact into the circulation for about 18 hours after birth. The mechanism involved in erythrocyte destruction in neonatal isoerythrolysis (NI) is a classic type 2 hypersensitivity reaction. When the antibodies specific for antigens present on the foal's erythrocytes are absorbed from the colostrum into the foal's bloodstream, they bind to the foal's erythrocytes. These antibodies can either opsonize red blood cells for removal by phagocytes, or can fix complement and lyse them.

2. Horses do not have naturally occurring antibodies against the erythrocyte antigens involved in NI (in contrast to the situation with the ABO blood type in humans). Therefore exposure is required to stimulate production. This can happen during parturition, as a small amount of foal blood often gains access to the maternal circulation. This is sufficient to sensitize the mare if the foal has a dominant erythrocyte antigen that is lacking in the mare. Subsequent foals that receive colostrum from the mare are then at risk of developing NI if they also carry the dominant erythrocyte antigen.

3. Icterus is a yellowing of the skin and sclera that is caused by accumulation of excess bilirubin. In this case the bilirubin is formed as a result of the breakdown of hemoglobin from the erythrocytes that have been destroyed by the immune reaction. The icterus index is a measure of the total bilirubin in the blood.

CASE 19

1. The loss of erythrocytes opsonized with IgG antibodies occurs through erythrophagocytosis in the spleen. Splenectomy is sometimes performed to decrease this removal of erythrocytes from the blood vascular system.

2. A primary idiopathic IMHA does not have an identified cause. The probable cause is a loss of tolerance to self antigens with precipitating causes, such as stress, polyclonal activation by disease, or vaccination, but such causes are not identified. In contrast, the secondary forms of IMHA may be caused by a drug metabolite acting as a hapten and initiating an immune response to an altered erythrocyte antigen, or by infection with one of several bacteria (such as tick-borne diseases). The secondary IMHA disease is easier to treat because once the inciting cause has been removed (for example, by stopping a drug that is a causal factor, or treating the primary disease), the anemia will resolve. In contrast, the idiopathic forms of the disease often require long-term treatment with an immunosuppressive drug.

3. The IgM antibody, being pentavalent, potentially has 10 antigen-binding sites per molecule. As such, only one molecule of IgM is needed to bind to adjacent antigenic epitopes on erythrocytes and to fix C1q of complement. This initiates the classical pathway of complement fixation, which terminates in membrane disruption and erythrocyte lysis. In contrast, two IgG molecules bound close together on a cell membrane are needed to cause C1q binding and initiation of complement fixation. Thus extravascular hemolysis is associated with IgG, and intravascular hemolysis with IgM. Complement fixation also produces C3b, an opsonin, as a byproduct when C3 is split. This is important when the mechanism of erythrocyte loss involves phagocytosis of erythrocytes by the fixed phagocytes that line the sinuses in the liver and spleen. The C3b bound to the erythrocyte binds to the C3b receptor on subsequent removal.

4. Although there is no proven association between frequent vaccination and IMHA, it has been suggested that the practice of annual vaccination with multiple vaccines could potentially cause polyclonal B-cell activation, which in some dogs may stimulate production of autoantibodies against erythrocytes. In a dog with idiopathic IMHA it seems prudent to avoid unnecessary immune stimulation.

CASE 20

1. A low hematocrit could be the result of decreased erythrocyte production, but in this case it was caused by the destruction of erythrocytes (red blood cells), either by antibody/complement-mediated lysis or by opsonization using C3b receptors or antibody and removal by fixed phagocytes in the liver and spleen. There could also be some loss from bleeding due to the low thrombocyte count.

2. The petechiae and epistaxis were caused by an impaired blood-clotting mechanism. Petechiae occur when very small blood vessels in the skin are traumatized and blood leaks from them. With few to no platelets present, the ability of the blood to clot is severely compromised and any small injury can result in prolonged bleeding.

3. A direct Coombs test was used to evaluate peripheral blood erythrocytes for the presence of antibodies bound to their surface. This was done using a "Coombs reagent," which is an antiglobulin that binds to the antibodies on the erythrocytes, causing them to agglutinate. Erythrocytes without antibodies bound to the surface will not agglutinate and will test negative. The test used for platelets was an immunofluorescence test performed on

a bone-marrow biopsy. Fluorochrome-labeled anti-equine IgG detected megakaryocytes that had antibody bound to the surface. Another type of test that can be performed on peripheral blood uses flow cytometry to identify platelets that have antibody bound, but it was not used in this case.

4. For complement fixation by the classical pathway to occur there must be two IgG molecules close enough to each other on the cell membrane for C1q of complement to bind to the Fc pieces and initiate the cascade. If this does not occur, the antibodies can still opsonize and cause removal of the erythrocytes by phagocytes, but not lysis by complement.

CASE 21

1. Both myasthenia gravis (Case 22) and pemphigus foliaceus have a type II hypersensitivity mechanism of pathogenesis. This means that antibodies develop against self antigens on cells or cell-associated proteins (acetylcholine receptor in myasthenia gravis, and desmoglein-1 in pemphigus foliaceus), and then these antibodies damage the cells, causing their death. Complement-mediated lysis is also caused by the bound antibodies fixing complement on the target cell in both diseases.

2. Once the bound antibodies have fixed complement, the terminal pathway results in loss of osmotic stability of the cell. The cell then dies and sloughs off, leaving a blister formation. In addition, chemotactic molecules are released that attract neutrophils to the area.

3. When a patient is on immunosuppressive drugs, the immune system is less able to mount an immune response to new pathogens that enter the body. Even though the suppression is intended to target the autoimmune response that is responsible for the disease, the response to other antigens is similarly depressed. Therefore protection with antimicrobial agents is an important component of therapy.

CASE 22

1. Pyridostigmine is a cholinesterase inhibitor. Inhibition of cholinesterase prolongs the biological half-life of acetylcholine, so that the small amount which is present in the neuromuscular junction is able to persist. Acetylcholine is required for transmission of the nerve impulse across the junction to stimulate the muscle to contract.

2. Treatment with corticosteroid (or more potent immunosuppressive drugs) acts on the cells of the immune system to decrease the production of autoantibodies, thus halting progression of the destruction of the acetylcholine receptors.

3. The removal of protein from the plasma of a patient with circulating autoantibodies can be an effective way to remove the product of the aberrant immune response. By combining this therapy with an immunosuppressive drug, the autoimmune response can be slowed or stopped and the product removed to prevent further tissue destruction.

CASE 23

1. Bover has immune complex disease and might be expected to have kidney damage due to immune-mediated vasculitis affecting the kidney glomeruli. The absence of protein indicates that the disease has not yet affected his kidneys.

2. The low albumin/globulin (A/G) ratio in this case means that the globulin fraction of the serum is increased while the albumin fraction is normal. Increased gamma globulins, as seen in a polyclonal gammopathy, would be reflected in active germinal centers in the cortex of the lymph nodes, causing hyperplasia.

3. ANA refers to antinuclear antibodies that are present in the blood of the SLE patient. It is these antibodies that complex with nuclear antigens from the break-up of apoptotic cells. The assay is performed by making serial dilutions of patient serum, incubating on fixed HEp-2 cells, and then staining with anti-canine IgG. The nuclei of the cells will fluoresce if the serum contains antibodies that bind to nuclear antigens.

CASE 24

1. The type III hypersensitivity response that causes purpura hemorrhagica involves immune complexes fixing complement and lodging in the walls of small blood vessels throughout the body. Complement components C3a and C5a are anaphylatoxins, and can release histamine from local mast cells. Histamine causes vasodilation and increased vascular permeability. These same components are chemotactic for neutrophils. Movement of neutrophils to the area of immune-complex deposition causes them to release their destructive lysosomal enzymes into the tissue. These enzymes cause destruction of endothelial cells, bleeding, and leakage of fluid into the surrounding tissue.

2. The development of purpura hemorrhagica is due to the formation of small immune complexes that escape removal by the reticuloendothelial system. The size of the immune complexes varies depending on the ratio of antigen to antibody (Figure 24.1). Complexes that form at equivalence (when antigen and antibody ratios are close to 1:1) are quite large, but when there is an antibody excess or an antigen excess these intermediate size complexes can lodge in the walls of blood vessels and cause vasculitis. Thus it appears that the amount of available circulating antigen (M protein of *Streptococcus equi*) and the amount of antibody produced are the factors that determine whether there will be immune-complex deposition. Most horses clear the antigen efficiently with a moderate antibody response. Those that fail to clear the antigen despite a vigorous antibody response are more likely to develop purpura hemorrhagica.

3. In serum sickness, a bolus of a foreign protein is injected into an animal (for example, horse serum for tetanus prophylaxis is injected into an unimmunized human), and the proteins in the serum are gradually metabolized (the half-life of IgG in humans is 21 days). During that time an antibody response is initiated. When the antigen and antibody are at appropriate concentrations for the formation of immune complexes, they are deposited in small blood vessels in the body: the kidney glomerulus and the joints are affected in particular. The patient develops glomerulonephritis and arthritis as a result of the deposition of these complexes, but the disorder is temporary, because the antigen is ultimately all metabolized. In post-strangles purpura hemorrhagica, large amounts of antibodies are produced against the *Streptococcus equi* proteins, and these antibodies form immune complexes with soluble proteins from the bacteria. The immune complexes can circulate as described above for serum sickness, and can be deposited in the small blood vessels, causing complement fixation and inflammation.

CASE 25

1. A type I allergic response is caused by the antibody IgE, which binds to mast cells in the skin, and upon contact with allergen it stimulates mediator release from the mast cell. These mediators (for example, histamine) increase the permeability of blood vessels, and a local wheal occurs within minutes of antigen/allergen contact. In contrast, the contact allergy is caused by sensitized T lymphocytes which are activated by small molecules that bind to self proteins in the skin, and it takes several days for a lesion to become apparent. After antigen presentation to T lymphocytes by Langerhans cells in the skin, the TH1 cells release cytokines and chemokines that stimulate macrophage recruitment and activation. In addition, epithelial cells also secrete cytokines and chemokines, all of which foster a pro-inflammatory environment.

2. Some contact allergens that are lipid soluble (such as urushiol) also elicit $CD8^+$ T cells that can become directly cytotoxic to epithelial cells that have bound allergenic haptens.

3. A blood test for antibodies to a sensitizing chemical would not be useful for diagnosis of contact hypersensitivity because antibodies are not involved in the causation of this disorder. Type IV hypersensitivity is a T-cell-mediated reaction.

CASE 26

1. Antibody binding to thyroglobulin, T3, and T4 can inactivate or remove these hormones, thereby preventing them from reaching their target receptors on a variety of cells and organs. Lack of these hormonal signals can decrease metabolism in target organs.

2. The T lymphocytes that are specific for determinants on the cells of thyroid follicles can initiate cell death with granzyme/perforin systems, thereby decreasing the source of new thyroid hormone synthesis. In addition, production of cytokines, such as interferon γ, contributes to an inflammatory environment that leads to follicle destruction.

3. Thyroid hormones act on almost all of the cells of the body. They are essential for maintenance of the basal metabolic rate, for protein synthesis, for carbohydrate, fat, and protein metabolism, and for vitamin and energy metabolism.

CASE 27

1. A type III hypersensitivity response will involve the formation of immune complexes between IgG and antigen in the lung, with subsequent complement fixation and release of chemotactic factors that stimulate the accumulation of neutrophils. Tissue damage occurs as a result of neutrophil degranulation. The type I hypersensitivity mechanism requires the production of IgE antibodies, which bind to mast cells in the lung and release histamine and other mediators when antigen is cross-linked on the mast-cell surface by the binding of IgE to antigen. These mediators cause smooth muscle contraction and increased vascular permeability, which results in airway lumen reduction and increased mucus formation. Both of these immune mechanisms can cause clinical signs of respiratory dyspnea and cough.

2. There are reports of families of horses with an increased incidence of RAO. For these horses there are also some studies that indicate a lower than normal incidence of infection with helminths, such as strongyles. Since we know that the ability to mount a high IgE response is inherited in several species, and that IgE is the normal immune response mounted in order to destroy helminth parasites, it makes sense that horses with the genetic predisposition are very likely to have some type I hypersensitivity as part of the pathogenesis of their RAO.

3. Dexamethasone is a corticosteroid that has an immunosuppressive effect. It will reduce the influx of neutrophils into the lung and will down-regulate the production of prostaglandins and leukotrienes, eicosanoids that collectively cause bronchoconstriction, leukocyte chemotaxis, and modulation of the immune response. Decreasing these pro-inflammatory effects allows the lung to return toward a normal state of functioning.

CASE 28

1. Dogs that develop hypersensitivity to flea antigens become sensitized after frequent bites from the flea, in which salivary antigens enter the tissue and blood of the dog during consumption of the blood meal by the flea. As with other immune responses, antigen-presenting cells engulf the antigens and process them, ultimately making peptides from the antigens available for recognition by CD4$^+$ T cells in local lymph nodes. These T-helper type 2 cells provide co-stimulatory signals (binding of co-stimulatory molecules and production of cytokines, IL-4, IL-5, and IL-13) to B lymphocytes that bind flea allergens through the B-cell receptors. B-cell proliferation and differentiation result in the production of IgE-secreting plasma cells. IgE binds to mast cells in the skin, which degranulate when flea allergens infiltrate the skin. Dogs with type IV hypersensitivity also become sensitized to flea allergens, but the response is dominated by interferon γ and IL-12, causing the proliferation of mononuclear cells. In one study it was shown that some dogs develop a type I response and subsequently develop type IV reactivity, while others develop the two allergic responses concurrently.

2. The immediate response to introduction of allergen by a flea bite is the formation of erythema and a wheal. Release of histamine and other mast-cell mediators causes a pruritic response, which initiates self-trauma and ultimately the development of erosions and ulcers. The numbers of mast cells containing chymase and tryptase increase in the skin of dogs with flea allergy dermatitis (FAD), and decrease after challenge/degranulation. The type IV response involves accumulation of mononuclear cells at the lesion site, with chronic lesions appearing as firm erythematous lesions without the characteristic wheals of the type I response.

3. Traditional "allergy shots" that are used in atopic humans and dogs to modulate the immune response in order to decrease IgE production serve as the model for the FAD allergy shot. An attempt to decrease IgE responses depends on frequent injection of ever increasing doses of allergen. The aim of the second type of FAD vaccine discussed is to build IgG antibodies to flea hindgut allergens so that blood meals will contain antibodies that kill the flea hindgut cells, and ultimately the flea. Thus the latter vaccine results in decreased flea populations on the dog.

CASE 29

1. Even though a patient with multiple myeloma has a very high concentration of antibody, it is not a functional antibody, and the ability of the dog to respond to new infections is limited due to the presence of neoplastic cells in the bone marrow.

2. The monoclonal spike is caused by the proliferation of a clone of plasma cells all producing exactly the same antibody (with the same variable region and electrophoretic charge). Normally an immune response will consist of multiple different plasma cells each producing an antibody with a different variable region, and thus the electrophoretic charges will be slightly different and the γ region on the densitometry tracing is broad.

3. The anemia seen in conjunction with multiple myeloma is caused by the accumulation of neoplastic plasma cells in the bone marrow, leaving little room for the hematopoietic cells that are important for production of erythrocytes.

CASE 30

1. Feline infectious peritonitis (FIP) virus is formed in the enteric coronavirus-infected cat by mutation (in the *3c* gene). It is thought that cats which carry the mutation are influenced at a young age by environmental factors (for example, crowding and poor sanitation), and that if appropriate genetic factors are present the virus will propagate unchecked in the kitten, and inflammatory lesions will occur. Not all kittens with a mutation in the *3c* gene of the enteric coronavirus develop FIP.

2. Because infection with the feline enteric coronavirus is so common, most cats have some antibodies against the virus. Therefore an antibody titer alone is not helpful in diagnosis. Those FIP cases that are diagnosed by clinical signs, abdominal fluid analysis, the presence of anemia, and gammopathy will have a much higher titer on the indirect immunofluorescence test for FIP than normal cats. In these cases the serum titer is confirmatory, but is not diagnostic on its own. The presence of FIP virus in macrophages, determined by polymerase chain reaction (PCR), is more diagnostic than the serum titer because the enteric coronavirus should not be present in macrophages.

3. FIP virus causes immunosuppression; specifically, it causes a reduction in the numbers of natural killer (NK) cells and T-regulatory cells. In addition, in those cats that show antibody-dependent enhancement, the antibodies fail to neutralize the virus and instead facilitate the entry of virus into macrophages, which then transport the virus throughout the body. FIP virus also causes a polyclonal gammopathy—that is, a large amount of antibody that appears to be ineffective in providing protection against the virus.

CASE 31

1. The albumin/globulin (A/G) ratio is normally in the range 0.5–1.14. When the ratio is high it can indicate either underproduction of immunoglobulins, or excessive amounts of albumin (this is less likely). A low A/G ratio usually reflects an increased number of immunoglobulins, as is sometimes seen in autoimmune disease and chronic infection. A low level of albumin may also indicate protein loss associated with renal or liver disease.

Further analysis of serum electrophoresis protein profile results will provide additional information about the polyclonal or monoclonal nature of a gammopathy.

2. As an obligate intracellular organism in a monocytic cell, *E. canis* resides in the endosomes of the host phagocytic cells and causes a persistent infection. The role of the CD4+ T-helper type 1 cells in protection against infections by bacteria that are able to survive in the phagosome is well recognized. Production of interferon γ by these T cells enhances host immunity by activating the mononuclear cells that are infected. Studies have shown that early production of the cytokine interleukin 10 can depress interferon γ production and allow the organisms to continue to propagate in the phagosomal compartment.

3. These three diseases include an autoimmune disease (systemic lupus erythematosus, SLE), a chronic viral infection (feline infectious peritonitis, FIP), and a persistent subacute to chronic intracellular bacterial infection (canine monocytic ehrlichiosis). In the case of SLE there is a loss of self tolerance and ongoing production of antibodies to a variety of self molecules, including but not limited to nuclear components. This causes a polyclonal activation of B lymphocytes with production of many antibodies of varied specificities. FIP is caused by a coronavirus infection of mononuclear cells, which can also stimulate many different B cells with resultant high levels of gamma globulins. *E. canis* infection is somewhat unique in that as a subacute to chronic infectious disease it often stimulates a polyclonal gammopathy, yet it is sometimes associated with a monoclonal or mixed (restrictive oligoclonal) gammopathy.

INDEX